WAYS OF THINKING ABOUT GOD

The Bible, Philosophy, and Science

E. B. Self

© 2013

Published in the United States by Nurturing Faith Inc., Macon GA,
www.nurturingfaith.net.

Library of Congress Cataloging-in-Publication Data is available.

978-1-938514-22-7

All rights reserved. Printed in the United States of America

The Scripture quotations contained herein are from the New Revised Standard Version Bible, ©1989 by the Division of Christian Education of the National Council of the Churches of Christ in the U.S.A., and are used by permission. All rights reserved.

CONTENTS

Introduction v

PART 1
God and the Bible

    Chapter 1: Defining Qualities of God 3
    Chapter 2: Descriptive Titles of God 15
    Chapter 3: Divine Actions 23
    Chapter 4: God and Violence 33
    Chapter 5: God and Human Destiny 45

PART 2
God and Philosophy

    Chapter 6: Trying to Prove God's Existence 65
    Chapter 7: The Nature of God 77

PART 3
God and Science

    Chapter 8: God and Time 91
    Chapter 9: God and Life 103
    Chapter 10: God and Space 119

Author's Views 131
Bibliography 143

INTRODUCTION

Many years ago, when I went from being a student at Baylor University in Waco, Texas, to being a student at Yale Divinity School in New Haven, Connecticut, I heard various stories. One story was about a student at a theological seminary in the South and a student at a seminary in the North. The student from the South asked the student from the North if there was not a difference between the seminaries in the two parts of the country. The student from the South was probably thinking of the traditional categories of conservative and liberal. The student from the North somewhat surprisingly agreed that there was a difference. He said, "We talk about God and Christ. You talk about smoking and drinking."

When I went from being a student at Yale Divinity School to being a student at Vanderbilt University in Nashville, Tennessee, there was another story that impressed me. One of the Vanderbilt Divinity School professors, a Methodist minister, had written a book on *Should Christians Drink?* There seemed to be confirmation that I was indeed back in the South, but there is more to the story. Another Vanderbilt Divinity School faculty member reportedly said that he, too, might be writing a book. The title would be *Should Christians Think?*

You may or may not see humor in these stories. Whatever else may be said, they have helped me to consider some of my own beliefs:

1. There are many appropriate subjects for Christians to discuss. Smoking and drinking were not big issues at Yale Divinity School, at least not while I was there. I did hear often angry talk at YDS about the evils of war, poverty, and racial prejudice. I believe it is appropriate for Christians to talk about smoking, drinking, war, poverty, racial prejudice, and many other topics.
2. The most important subjects for Christians to consider are "God and Christ." Christians vary in which topics they find interesting and even exciting, but no other subjects are greater in importance than God and Christ. There would be no Christianity, at least not in the traditional sense, without them. We cannot make that statement for many other subjects.

3. Christians should not only emphasize talking about God and Christ but should also think carefully about these subjects. Talking, even talking vigorously, without thinking is not a good idea. The apostle Paul wrote about those who had great concern for God but were still lacking. He mentioned those who "have a zeal for God but it is not enlightened" (Rom. 10.2). However zealous we may be, we should all seek enlightenment about God and Christ. One way to gain enlightenment is through careful thinking.

As part of my thinking about Christ, I wrote a book on *What Should We Believe About Jesus?* My purpose was to consider what we should believe about Jesus according to the New Testament. Although I do not believe that every Christian needs to write a book about Jesus, all Christians should think carefully about him.

As part of my thinking about God, I wrote a book on *Loving God with Your Mind.* In the Gospel of Mark, a Jewish scribe asks Jesus the question of which commandment is greatest or "first of all." Jesus answers, "The first is, 'Hear, O Israel: the Lord our God, the Lord is one; you shall love the Lord your God with all your heart, and with all your soul, and with all your mind, and with all your strength'" (12.29-30). All the ways of loving God are important but are not always followed. Loving God with all your mind is one of those ways; it calls for careful thinking.

I wrote a book with attention to thinking about God biblically, philosophically, and personally (in relationship to persons). I would now like to expand that book and adjust the organization. I plan to consider three ways of thinking about God: (1) in relation to the Bible, (2) in relation to philosophy, and (3) in relation to science. I have decided to use the title of *Ways of Thinking about God: The Bible, Philosophy, and Science.*

Thinking about God in relation to the Bible includes numerous beliefs about him. How can the beliefs about God in the Bible be summarized? There are different ways. I have chosen to use the categories of defining qualities, descriptive titles, and divine actions. Among various special considerations about God and the Bible, I have selected God and violence and also God and human destiny.

There are many good translations of the Bible. In order to reduce complications, however, I have used one translation: the New Revised Standard Version as incorporated in the third edition of *The New Oxford Annotated Bible.* No translation is perfect, but the NRSV has a clear style and a high reputation for scholarship. Also, the NRSV provides some alternative translations.

Thinking about God in relation to philosophy involves an emphasis on human reason, usually reason apart from special revelation such as the Bible. While there can be reasoning about God as presented in the Bible, philosophers often reason about God without much reference to the Bible. Philosophers have often tried, for example, to reason about God from the world of nature.

For those who are not familiar with philosophy, there is basic information in such introductory books as *Invitation to Philosophy* by Stanley Honer, Thomas Hunt, and Dennis Okholm. The authors tell us that philosophy in the western part of the world began in ancient Greece (several centuries before the time of Jesus). The original meaning of philosophy was "love of wisdom." There have come to be additional meanings, including a rigorous, rational attempt to understand. When philosophers attempt to understand anything, including God, they emphasize careful thinking. Their efforts include such basic questions as "What do you mean?" and "Why do you think that?"

While philosophers have written vast amounts about God, I have chosen to concentrate on two important topics. One philosophical topic is trying to prove the existence of God by reason apart from the Bible. Philosophers who have believed in God have disagreed over whether such proof is possible. The other selected topic for God and philosophy is the nature of God. Many philosophers who have believed in God have accepted traditional understandings of God's nature. Some philosophers have had special views of God's nature. Also, there have been differences between philosophers over the claimed goodness of God, especially in relation to evil in the world.

Thinking about God in relation to science is another way of thinking about God. Or is it? Is God an appropriate subject for science?

While there can be complicated explanations, science is basically knowledge gained from the senses. The emphasis in the scientific method is on knowing by observation and by reasoning on the basis of what is observed. The observation comes from the five physical senses of sight, hearing, touch, taste, and smell. The senses may be aided by special instruments such as microscopes, telescopes, spectrometers, and computers. Scientists believe that it is by such observation that we can know reality.

If we think of God as Spirit as Jesus taught, we cannot observe him by the physical senses. How then can scientists use the scientific method to study God? Many scientists have concluded that they cannot observe God, even with special instruments. Thus it seems that God is not a proper subject for scientific investigation. Although some scientists are persuaded that God is the invisible power behind the visible world, scientists generally regard that view as a religious belief and not a scientific one.

There may still be an appropriate way for thinking about God in relation to science. We can consider some traditional beliefs about God in view of various scientific discoveries and accounts. I wish to follow this approach in three selected areas: time, life, and space. When we think of traditional views about God, what views are there about him in relation to time, life, and space? When we think of science, what views have developed about time, life, and space? To what extent do the views agree or disagree? Since I am not a scientist, I will simply do my best to understand some of the main beliefs of scientists and how those views may influence traditional beliefs about God.

As we become more deeply involved in thinking about God, it would be good to remember that the thinking is about God but is done by humans. We should be careful. We should not have too high an opinion of our own thinking. We might recall the statement of Isaiah 55.8-9: "For my thoughts are not your thoughts, nor are your ways my ways, says the Lord. For as the heavens are higher than the earth, so are my ways higher than your ways and my thoughts than your thoughts."

With the reminder that human thinking about God has important limitations, we should still try to think about God as carefully as possible.

PART 1

***

God and the Bible

CHAPTER 1

Defining Qualities of God

An important way of thinking about God involves studying the Bible, a way that many people consider not only the most important way but also perhaps the only proper way. What did the biblical writers say about God? What do we think about what they said?

A famous verse in the Bible expresses the view that God is but gives no further indication of what God is. In the story of the call issued to Moses, there is this account: "God said to Moses, 'I AM WHO I AM.' He said further, 'Thus you shall say to the Israelites, "I AM has sent me to you" ' " (Exod. 23.14). This verse is profound, but what do the biblical writers claim about God in addition to his existing? We usually like an expression such as "I am" to be followed with more information.

The Bible does not contain a dictionary definition of God or a complete description of all that God is. The Bible does have what may be called defining qualities of God. These qualities are expressed in various places that have the idea of "God is." The verses often have little words with great significance. In the Old Testament there are statements that God is one, God is good, and God is holy. In the New Testament there are assertions that God is perfect and that he is spirit, light, and love.

God Is One

According to the NRSV translators, the writer of Deuteronomy 6.4 stated: "Hear, O Israel, the LORD is our God, the Lord alone." Jews often refer to this verse as the Shema, the Hebrew word for "hear," which is the first word of the statement. Jews and Christians think of this verse as an indication of belief in one God. That meaning is more clearly stated in the three alternative translations given in the NRSV. One of the alternative translations is stated in this way: "The LORD our God, the LORD is one."

Monotheism is the term for belief in one God, whereas polytheism designates belief in many gods. There are representatives of both views.

Except for a brief period of monotheism under Akhenaton (around 1375-1358 B.C.E.), the ancient Egyptians were polytheists. If we count minor deities, the Egyptians worshipped many gods and goddesses. Included were Osiris, Isis, Seth, and Horus. A strange feature is that some of the Egyptian deities were represented as having human bodies with the heads of animals. The ancient Hebrews were probably very familiar with Egyptian polytheism, because they lived in Egypt for several centuries between the times of Joseph and Moses.

The classical civilizations of ancient Greece and Rome officially followed polytheism. The Romans accepted the main Greek gods but gave them different names. If we take into account minor gods and spirits, the ancient Greeks recognized hundreds of deities. The main Greek gods were the Olympians, who supposedly lived on Mount Olympus and had games played in their honor. Some of the Olympian gods (and their Roman counterparts) were Zeus (Jupiter or Jove), Hera (Juno), Poseidon (Neptune), Demeter (Ceres), Athena (Minerva), and Dionysus (Bacchus).

In the first part of the second century B.C.E., Antiochus IV (Epiphanes) tried to force Greek culture—including the Greek gods—on the Jews. The Maccabees helped to gain independence for the monotheistic Jews for about a century before the Romans took control of Jewish land. The Romans were dominant both before and after the birth of Jesus and added emperor worship to their polytheism.

Hinduism, the major religion of India, has included polytheism for many centuries. Prominent in early Hinduism were the gods Indra, Agni, Varuna, and Soma. Then Vishnu and Shiva, along with Brahma, gained attention. Vishnu's claimed avatars or incarnations include Rama and Krishna. There are many other gods and goddesses in Hinduism even today.

Shinto, the native religion of Japan, is also polytheistic. There has been belief in myriads of kami or spirits. Many Japanese believed their emperors were descendants of kami. During World War II, Japanese commanders had kamikaze pilots, who went on suicide missions in honor of their beloved emperor. After the war ended in 1945 and Japan was occupied by allied forces under General Douglas MacArthur, Emperor Hirohito was required to admit that he was not divine. Japanese belief in kami is apparently not as strong as it once was.

Although polytheism was strong in the ancient world and still has followers today, many people have followed monotheism. In spite of some strong disagreements in other areas, Muslims agree with Jews in believing in one God. Muslims have in their confession of faith that there is no god but

Allah. For Muslims, Allah has no partners. Sikhism, a religion found mostly in India, received strong influence from Islam. The followers of Sikhism believe there is only one God.

Christians agree with Jews and Muslims in claiming monotheism, but Christians have a special idea of the oneness of God. The Christian doctrine of the Trinity is that there is one God in three persons. The persons are Father, Son, and Holy Spirit. The doctrine of the Trinity may look like polytheism to Jews and Muslims. The belief of Jews and Muslims in one God is comparatively simple. How can Christians have the unusual interpretation that God is both one and three? Traditional Christians accept the doctrine of the Trinity as both a profound truth and a mystery.

The biblical teaching that God is one remains a highly significant belief about God.

God Is Good

The Bible also says that God is good. The psalmist made this declaration in Psalm 106.1: "O give thanks to the Lord, for he is good." (See also 1 Chron. 16.34.) In the New Testament, in Luke 18.19, Jesus is reported to have said, "No one is good but God alone."

How should we understand the belief that God is good? When we refer to humans as good, we often think of not doing anything bad. A person may be good by not breaking laws and not participating in vices. But being good by entirely avoiding anything bad does not seem to be possible for humans. We all do or think things we should not do or think. Perhaps this situation is what Jesus was thinking when he said that God alone is good. Believers cannot imagine that God would do or even think anything bad.

But how do we deal with biblical statements that indicate God as doing things we would ordinarily consider to be bad? We find places in the Old Testament especially where bad consequences are attributed to God. Consider, for example, a part of Amos 3.6: "Does disaster befall a city, unless the Lord has done it?" Disaster for a city does not sound good. It sounds terrible for the people who live there. How could disaster for a city come from a good God? In order to keep thinking of God as good, we would have to believe that God is good no matter what God does. Perhaps things that may seem bad to us are part of God's overall good plan, a plan we may not fully understand. Might God show goodness by properly punishing people who do evil? Does God sometimes bring bad consequences in order to teach a good lesson?

Another way of thinking about being good is to meet standards of goodness. There are various moral principles that provide standards. One moral principle is fairness. We usually think that anyone who is fair, including God, is good. Another moral principle is mercy. We usually think that anyone who is merciful, including God, is good. If God is both fair and merciful, surely we could consider God to be good.

We can encounter complications. Think of those two moral principles of fairness and mercy. The two do not always go together. Would not God be fair, but not merciful, if God punished someone in direct proportion to the person's sin? Fairness requires giving what is deserved. In this case, God would meet the standard of fairness but not the standard of mercy. But God might forgive a misdeed and spare a person from punishment. That action would not be strictly fair of God because the person did not get what was deserved. Yet the guilty person would probably consider God to be good for showing mercy. We may well wonder if there is a way to combine fairness and mercy when we think of God's goodness. The difficulty may be more of a problem for our limited thinking than for God's goodness.

Yet another way of thinking about being good is to provide benefit. The emphasis here is not on the absence of something bad but on the presence of something positive. We usually think of someone as good if that person provides a worthwhile gift or valuable service for someone else.

Is God good in the sense of providing benefit? We find a positive answer to this question in the Psalms. The writer of Psalm 106.1 explained his view that God is good by saying "for his steadfast love endures forever." Whatever else may be said about God's goodness, the psalmist associated it with the great gift of God's love.

God Is Holy

We further find in the Bible the belief that God is holy. In Isaiah's vision in the year that King Uzziah died, one of the seraphim said, "Holy, holy, holy is the LORD of hosts; the whole earth is full of his glory" (Isa. 6.3). (See also Lev. 11.44-45 and Isa. 30.15.)

What is meant by thinking of God as holy? Isaiah experienced a deep sense of the power of God. With the conviction that God's glory filled the whole earth, there is an indication of tremendous might. We do find indications in the Old Testament that God's holy power may be concentrated in a particular place, such as the Ark of the Covenant or the room in the temple designated as the Holy of Holies. But in Isaiah 6 there is the view that

God's holy power extends throughout the whole earth. Isaiah probably felt that God as holy deserved immense respect, reverence, and awe.

Isaiah also was conscious of his own unworthiness in the presence of moral superiority. According to Isaiah 6.5, the prophet said, "Woe is me! I am lost, for I am a man of unclean lips, and I live among a people of unclean lips." The sense here seems to be much more than a feeling of ritual impurity. Isaiah's remorse came from a profound experience of the Holy One, who had high requirements for proper behavior. The prophet had a keen sense of his own failings when he thought of the moral excellence required by God's holiness. Fortunately for Isaiah, the prophet also had a sense of forgiveness. Isaiah's vision included a cleansing touch of his mouth with a live coal from the altar.

Among Christians there have been different responses to the idea of God's holy power. Some Christians have concentrated on God's awe-inspiring greatness. There has been expression of reverence for God's holiness in formal and dignified services of worship. The idea is humble adoration of the magnificent glory of God. Some Christians believe that respect for God's holiness should not lead to a sense of distance from God. These Christians do not deny God's holiness but seek a sense of personal closeness to God. Their worship services tend to be informal with more emphasis on how God mercifully cares for them in spite of their failings.

There have also been different responses to the moral part of God's holiness. Some Christians have emphasized what they believe are God's holy and strict standards for personal behavior. Failure to live up to those standards would result in feelings of great shame and would perhaps bring exclusion from the company of those with better conduct. Some other Christians have not rejected moral standards but have emphasized God's forgiveness when expected lapses from good behavior do occur. There is the example of Isaiah, who had a deep sense of God's high standards but also had a sense of cleansing.

Holiness is a profound and sometimes difficult concept. Christians agree that God is holy but have different interpretations of how God's holiness should influence their lives.

God Is Perfect

When we go from the Old Testament to the New Testament, we find additional defining qualities of God, including being perfect. According to Matthew 5.48, Jesus taught, "Be perfect, therefore, as your heavenly Father is perfect."

It is not startling to think of God as perfect. We would not expect anything other than the highest for God. But how can humans attain what seems to be an impossible standard? Did not Jesus teach in Luke 18.19 that "No one is good but God alone"? What does it mean to be perfect?

One meaning of being perfect is having no defect, fault, or error. That meaning sounds appropriate for God. How could he be anything less? But what is our situation as humans? Is any human without defect, fault, or error? The case for human imperfection is overwhelming. Yet God's perfection presents the ideal for what people should be. However defective we are, the challenge is to improve toward a goal that may be impossible but is inescapable.

Another meaning of being perfect is being complete. There is nothing missing. Everything that should be present is there. Can we think of anything missing with God? There is nothing lacking with him. The condition of humans is vastly different. No one is complete. There is something missing with everyone. It is not a reality but a goal for humans to be complete as God is complete.

John Wesley, the founder of Methodism, gave an important expression that applies to these meanings of perfection. He spoke of "going on to perfection." Many Christians can accept the view that they are not now perfect but could ultimately be made perfect by God's power. Most, perhaps all, would probably say that being without fault and being complete will not occur until a time beyond this life.

If we look at what preceded Jesus' teaching about perfection in Matthew 5, we find that his words may have had special meaning for the near future and even the present. How are Jesus' followers to be perfect as God is perfect? Jesus had spoken of loving even one's enemies. Did not God make the sun rise and send rain for both the righteous and the unrighteous? We are imperfect, but we should be more like God in how we provide for all people. This thought may have been what Jesus wished to emphasize in saying that his hearers should be perfect as God is perfect.

God Is Spirit

In the New Testament, God is also described as Spirit. According to John 4.24, Jesus said, "God is spirit, and those who worship him must worship in spirit and truth."

How should we understand God as spirit? Philosophical materialists think that only material things are real. If God is spirit and not material, then materialists believe God is not real. Also, there are those who believe that

God as spirit is real but have difficulty in trying to understand what is meant. Someone said that the best he could do with the view of God as spirit was to think of an oblong blur.

We may find it helpful to think of God as spirit if we consider various realities that are invisible. Oxygen, for example, is a colorless gas but is necessary for human life. We do not have to see oxygen in order to breathe it. Electricity is a powerful force but is usually unseen. Unless there are sparks, we cannot tell by looking whether or not a wire intended to carry electricity is actually doing so. Someone who does not believe a wire is live because he sees nothing might get a shock. Magnetism and gravity are additional unseen realities. No one has seen magnetism or gravity as such, but there is little doubt that they are actual forces. These examples do not prove the existence of God as spirit, but they do show that we often accept the reality of some things we cannot see.

It may also be helpful to consider that the Greek word used for God as spirit is *pneuma*, one meaning of which is "breath." The relationship between breath and human life is crucial. A person must have breath to live. Think of the biblical story of creation in Genesis 2. The writer stated in verse 7 that "then the LORD formed man from the dust of the ground and breathed into his nostrils the breath of life; and the man became a living being." Thus many believe that God as spirit is the source of breath and the life that goes with it.

Another meaning of the Greek word for spirit is wind. The wind itself is not visible and is hard to explain. It varies from a gentle breeze to a powerful tornado or hurricane. Whatever else we may consider, wind represents mysterious power. The Greek word for spirit was used in John's account of the conversation between Jesus and Nicodemus. According to John 3.8, Jesus said to Nicodemus, "The wind blows where it chooses, and you hear the sound of it, but you do not know where it comes from or where it goes. So it is with everyone who is born of the Spirit." The teaching seems to be that God as spirit provides mysterious power, somewhat similar to wind, in the lives of people.

If we think of God as spirit, associated with both breath and wind, is it appropriate to try to represent God in material form? Jews and Muslims do not picture God in art. The omission is not a matter of opposition to art itself. Jews have highly developed religious art in such expressions as ornate Torah scrolls. Muslims are well known for arabesque designs and stunning architecture for mosques. Jews and Muslims do not approve any artistic images of God because of opposition to idolatry.

Christians have been divided over God in art. Some see great value in pictorial expressions. For example, there is appreciation of Michelangelo's famous painting of God as creator in the Sistine Chapel in Rome. The magnificent work shows God with the graying head of an older man, perhaps suggestive of wisdom. The body of God in the painting is very muscular, no doubt indicative of power. Part of the explanation for this kind of portrayal of God is the view that many people are literal-minded and need to visualize God. A theological explanation for depicting God in visible form is Christian belief in the incarnation. With the conviction that God was present in the body of Jesus Christ, there is the idea that artistic representations of God in material form are both acceptable and helpful.

Other Christians have disagreed. Like the Jews and Muslims, they have been much stricter. The reasoning is not only opposition to idolatry but also the thought that portraying God in art is inconsistent with a literal understanding of God as spirit. Is not spirit invisible?

It is true that God as spirit without further associations is a difficult concept. Thinking of God as an oblong blur does not help much. The additional meanings of breath and wind for spirit do provide some help. We can think of God as spirit, as an invisible and mysterious but benevolent power that is crucial for human life and human fulfillment.

God Is Light

In the New Testament we also find references to God as light. The writer of 1 John 1.5 said that "God is light and in him there is no darkness at all." We find the same thought in Revelation 21.23: "And the city has no need of sun or moon to shine on it, for the glory of God is its light, and its lamp is the lamb."

There are some questions of interpretation. Why did the writer of Genesis 1.3 say that God created light? If God is light, why would God create it? What kind of light was this created light? The light of Genesis 1.3 on the first day of creation was not the light of the sun. According to Genesis 1.16, God did not create "the greater light to rule the day" until the fourth day of creation. Thus we have God as light, the sun as light on the fourth day of creation, and another kind of light on the first day of creation. The biblical view is that God created the other two kinds of light, but how are we to understand God as light in relation to them? Perhaps the other two kinds of light should be understood as physical light with God as light that exceeds physical limitations.

Another question of interpretation is why different ones are all said to be light. The claim of the writer of 1 John 1.5 is that God is light. In the Sermon on the Mount, Jesus referred to others as light. In Matthew 5.14, Jesus said, "You are the light of the world." It is not clear how many people he was addressing, but his disciples were included. (See Matt. 5.1-2.) Jesus was probably speaking here in a symbolic, primarily spiritual, sense. Jesus also spoke of himself as light. In John 8.12a, Jesus said, "I am the light of the world." Thus the Bible contains references to God, others, and Jesus as light. What is the proper interpretation of these verses about different ones being referred to as light? One possibility for the disciples (and other followers of Jesus) is to consider that the light of the moon does not originate with the moon but is reflected light from the sun. Perhaps we should think of the disciples (and other followers of Jesus) as reflected light.

What is meant by thinking of God as light? It probably would not be helpful in understanding God to investigate light as a form of electromagnetic radiation. It may be helpful to consider how we often associate light with other qualities. We may, for example, associate light with goodness. We associate darkness, which is the absence of light, with evil. To say that God is light may be a way of referring to God's goodness, a quality we have previously considered.

We also associate light with knowledge, while darkness represents ignorance. How much is God supposed to know? Theologians speak of God's omniscience, that is, God's power to know everything.

Life is a further association we have for light. As reported in John 8.12, Jesus said that those who follow him "will never walk in darkness but will have the light of life." Most plants and trees could not live without the action of light in the process called photosynthesis. The sun is literally the light of life for virtually all members of the plant kingdom. One interpretation of God as light is that he is the spiritual light of life for humans.

However the questions of interpretation are answered, there are strong associations of God as light. These associations include goodness, knowledge, and life. God as light is a powerful symbol for various beliefs about God.

God Is Love

A highly regarded belief about God is that he is love. The author of 1 John 4.8 stated, "Whoever does not love does not know God, for God is love." Greek words for love include *eros* (romantic love) and *philia* (brotherly love). The Greek word for love in 1 John 4.8 is *agape*, which is unselfish and giving love. The meaning of God as love is that God has an abundant and overflowing concern for the welfare of others.

Some verses in the Old Testament do not contain the statement that God is love but do refer to God's love in strong ways. Several verses include the Hebrew word *hesed*, meaning steadfast love. There is this statement in Psalm 51.1: "Have mercy on me, O God, according to your steadfast love." The writer of Psalm 89.1 declared, "I will sing of your steadfast love, O LORD, forever." We read this admonition in Psalm 106.1: "Praise the LORD! O give thanks to the LORD, for he is good; for his steadfast love endures forever." Jeremiah referred to God's "everlasting love" (Jer. 31.3).

Some New Testament verses refer to God's love with powerful claims. According to the teaching of Jesus in John 3.16, "God so loved the world that he gave his only Son, so that everyone who believes in him may not perish but may have eternal life." Paul proclaimed in Romans 5.8, "God proves his love for us in that while we still were sinners Christ died for us."

Although God as love is one of the most appealing qualities of God we have studied, not everyone has agreed with or emphasized this belief. The deists, who were strongest in the seventeenth and eighteenth centuries, thought of God as indifferent. They thought that God created the world and then had nothing more to do with it, letting it proceed on its own way. In colonial America, Jonathan Edwards thought that God was interested in people but was greatly upset with them. Edwards, whatever else he may have believed, preached a famous sermon on "Sinners in the Hands of an Angry God." Many of us have heard evangelists who, often in a hateful way, emphasized the belief that an offended God had prepared eternal damnation for unrepentant people.

The Bible does contain references to God's anger as well as to God's love. We might wonder if it is loving people who emphasize God's love while angry people emphasize God's anger. Whatever people tend to emphasize in their beliefs about God, a careful approach to biblical interpretation will include consideration of God as love without overlooking the references to God's anger. We may hope and believe that the love is greater than the anger.

Conclusion

The "God is" verses are a good place to begin when we think about God in relation to the Bible. Defining qualities of God in the Old Testament include references to God as one, good, and holy. The New Testament has statements of God as perfect, as spirit, as light, and as love. If we try to combine the qualities, we might say that God appears in the Bible as the one, good, holy, and perfect Spirit who is light and love. The combination may be accurate, but the qualities may be more impressive when stated singly.

There are additional possibilities in the Bible for defining qualities of God. Although not in the "God is" form, there are suggestions in the Bible that include God's eternity, changelessness, omnipresence, and infinity. Without considering everything, we have covered some important biblical qualities of God.

While defining qualities are a good place to start, the Bible does have much more information for thinking about God.

Questions for Further Thought

1. Why do you think that the "God is" verses should or should not be considered equal in importance?
2. Why do you think that additional defining qualities of God do or do not deserve as much consideration as the ones mentioned in this chapter?
3. To what extent do you think that their own personalities may influence how people differ on which qualities of God they emphasize? (For example, do loving people tend to emphasize God's love?)
4. Although the Bible has separate defining qualities for God, why do you think the Bible does not have a complete definition or description of God in one place?

CHAPTER 2

Descriptive Titles of God

In addition to defining qualities, the Bible contains descriptive titles for God. The Old Testament has many titles for God. The list includes creator, judge, redeemer, shepherd, king, husband, and father. Most of these titles appear in the New Testament, too. How should we interpret these titles? What do the biblical titles for God indicate about him?

Creator

God's activity of creation appears in the first two chapters of the Bible. The title of creator occurs later. The writer of Ecclesiastes 12.1 gave this admonition: "Remember your creator in the days of your youth." The book of Isaiah has questions and an answer: "Have you not known? Have you not heard? The LORD is the everlasting God, the Creator of the ends of the earth" (Isa. 40.28). In the New Testament, Paul wrote about those who have "exchanged the truth about God for a lie and worshiped and served the creature rather than the Creator" (Rom. 1.25). The author of 1 Peter 4.19 urged, "Therefore, let those suffering in accordance with God's will entrust themselves to a faithful Creator, while continuing to do good."

One view of the biblical writers is that God is responsible for the existence of the world. The traditional interpretation is that God created *ex nihilo*, which means out of or from nothing. According to this interpretation, God did not shape anything already existing but brought everything into existence.

Belief in God as creator out of nothing conflicts with a scientific principle I heard long ago: that matter can be neither created nor destroyed. Matter, so the explanation went, can change form. Ice, for example, can melt into water through the application of heat. With even more heat, the resulting liquid can then evaporate into water vapor or steam. But the idea was that you cannot destroy matter itself. And you cannot bring it into existence. An implication of this view is that matter of some kind has always existed and will exist forever, even if not in its present form.

For those who believe in God as creator, the understanding of matter as eternal is not accurate. The view of traditional believers is that matter has not always existed. God alone can create matter and did create matter and presumably can destroy matter. Although humans can alter matter, they cannot create it or destroy the matter God has made.

The biblical title of God as creator helps express the belief that God is responsible for the existence of both the world and people. If God is ultimately responsible for the existence of people, then humans are related to tremendous power of cosmic significance. If God is not the creator of people, then human importance is greatly diminished. The title of God as creator is important both for belief about God and for belief about the relative significance of people.

Judge

With the belief in God as creator, it is easy to think further that God must be in charge of creation. The people of creation would be accountable to God. Thus God as judge has a basis in God as creator.

There are various references in the Bible to God as judge. We read in Psalm 50.6: "The heavens declare his righteousness, for God himself is judge." The reference in Psalm 94.2 is to God as "judge of the earth." The writer of Hebrews 12.23 spoke of God as "the judge of all." And what kind of judge would God be? There is the famous question asked by Abraham in Genesis 18.25: "Shall not the judge of all the earth do what is just?"

Several New Testament references present Jesus Christ, the Son of God, as judge. In John 5.22 we read that Jesus said, "The Father judges no one but has given all judgment to the Son." One of Peter's speeches in Acts contains this statement about Jesus Christ: "He commanded us to preach to the people and to testify that he is the one ordained by God as judge of the living and the dead" (10.42). The writer of 2 Timothy 4.1 declared that it is Christ Jesus "who is to judge the living and the dead."

Why would Jesus Christ, the Son, be the judge? A possible explanation is the belief that the Son as the Word was the agent of creation for the world. (See John 1.1-3 and Heb. 1.1-2.) It would then be appropriate for the Son to judge at the final accounting.

The title of judge, whether Father or Son, can be appealing when we think of other people. We might be pleased to consider divine judgment of the wicked. We probably all know of people we would like to see punished. It can be irritating that some humans apparently get by with all kinds of outrageous behavior. Those people should at least eventually get the condemnation they deserve.

But what if we are the ones being judged? What if we learned that the judge would be completely fair? That situation might not be comforting. The psalmist asked, "If you, O LORD, should mark iniquities, Lord, who could stand?" (Ps. 130.3). We might feel like Isaiah. He is the one who said, "Woe is me! I am lost, for I am a man of unclean lips" (Isa. 6.5). However much we might like a strictly fair judge for others, we might not like such a judge for ourselves. We might prefer to be treated with abundant mercy rather than be judged with complete fairness.

Whether we like the idea of God as judge or feel some discomfort with it, there are clear indications of the title in the Bible. Whether the judge is God the Father or the Son, the biblical message is that we will be judged.

Redeemer

The title of redeemer for God helps lessen the possible strictness people may face with an entirely fair judge. If we have done wrong, going before a judge does not sound good. What if there is someone who can help us?

Job said, "For I know that my Redeemer lives, and that at the last he will stand upon the earth" (Job 19.25). The psalmist requested, "Let the words of my mouth and the meditation of my heart be acceptable to you, O LORD, my rock and my redeemer" (Ps. 19.14). The writer of Psalm 78.35 stated, "They remembered that God was their rock, the Most High God their redeemer." The book of Isaiah has several references to God as redeemer. (See 47.4, 49.26, 60.16, and 63.15.)

The basic meaning of a redeemer is one who buys back. Someone might buy back land, especially land that was lost through earlier failure to pay for it. A redeemer might relieve a debtor by paying what is owed. There could be redemption of a person from slavery by paying the required price. A redeemer is one who provides some kind of help for a person in need.

What is the meaning of God as redeemer? We do not find a complete explanation in the Bible. God does not pay money for anything, so what does God do? One answer is that God is able to help those who are in spiritual trouble and does provide that help. Perhaps a person has lost a close relationship to God. There is the possibility of redemption of that relationship through God's resources. Perhaps a person has a spiritual debt because of sin. God as redeemer can provide the means to deal with that problem. Maybe someone has become enslaved to an evil force. Then God as redeemer may bring deliverance. The overall view is that God as redeemer can somehow make things right, perhaps restoring what had been good but had been lost.

God's title of redeemer in the Old Testament is not fully developed and allows for various interpretations. A further thought is that God expressed his greatest redemptive work through Jesus Christ.

Shepherd

There are references in both the Old Testament and the New Testament that use the title of shepherd for God or Christ.

A well-known statement in the Old Testament is Psalm 23.1: "The LORD is my shepherd, I shall not want." The writer of Isaiah 40.11 said of God, "He will feed his flock like a shepherd; he will gather the lambs in his arms, and carry them in his bosom, and gently lead the mother sheep."

The New Testament has references to Jesus as a shepherd. In John 10.11 Jesus said, "I am the good shepherd. The good shepherd lays down his life for the sheep." Hebrews 13.20 refers to "the God of peace, who brought back from the dead our Lord Jesus Christ, the great shepherd of the sheep." We read in 1 Peter 5.4 of "the chief shepherd."

Most people today are not as familiar with shepherds as people were in biblical times. In spite of the passage of centuries, a shepherd's provision for his sheep is a clear example of God's provision for his people.

How much provision do sheep need? The animals are famous for lack of intelligence, as in the expression "dumb as sheep." They have little sense of direction. They are easily lost if they stray from the flock. They need almost constant care and supervision. It is not flattering to think that we as humans may be like sheep. We do not believe we are as stupid as they are. Yet we have to admit that we do need food and safety and we sometimes stray spiritually.

The title of shepherd for God is not something we take literally. We do not think of God as standing in a meadow somewhere with animals around him. But the idea of God's provision for his people, as a shepherd cares for his sheep, is compelling.

King

It is difficult to determine exactly when various titles for God first appeared. Since the Hebrews kept sheep long ago, the title of shepherd for God may be very early. The title of king for God probably came later, perhaps after the Israelites had kings of their own.

In his famous vision, Isaiah saw the LORD sitting on a throne (Isa. 6.1) and said in 6.5 that "my eyes have seen the King, the LORD of hosts." The Psalms contain several references to God as king. Psalm 5.2, for example, has this statement: "Listen to the sound of my cry, my King and my God, for to you I pray." (See also Ps. 10.16, 74.12, and 84.3.) Paul provided inspiration for a hymn known by many Christians: "To the King of the ages, immortal, invisible, the only God, be honor and glory forever and ever" (1 Tim. 1.17).

John Calvin (1509-1564) was deeply impressed with the idea of God as king. Calvin decided to emphasize God's sovereignty as the main theme for his theological system, views expressed in his *Institutes of the Christian Religion*.

Calvin thought of God as the absolute monarch for the world. God, Calvin believed, is all-powerful and is behind everything that happens. God as sovereign may do anything he wishes. We should not question God in any way. We may not fully understand, but God knows what is best. What if God does something that looks bad? Calvin's explanation was that humans do bad things for bad purposes while God does apparently bad things for good purposes. What if a tree falls in the forest and kills a man? Calvin thought that God was ultimately responsible not only for that action but also for everything that happens. And, Calvin thought, whatever is done by God is good and right.

Calvin's ideas about God are similar to the political philosophy of the divine right of kings—that kings have the power and right to do whatever they wish. A problem is that many human kings have done some very bad things. There are numerous accounts of bad kings in the Old Testament books of 1 and 2 Kings. Surely God is not like those kings and other horrible kings throughout history. Consider Jehoram, who reigned in Jerusalem. According to the account in 1 Chronicles 21.1-4, Jehoram killed all of his brothers when he became king. He wanted to remove rivals for the throne. It does not sound at all appropriate to think of God as being like Jehoram or any other cruel and barbaric king.

It was probably not Calvin's intention to think of God as being like a bad king, except for the idea of power that all kings have. An important difference for Calvin was that bad kings do truly bad things while God may do something that looks bad but is actually good. For Calvin and his followers, we should not think that God somehow may have the defects of bad kings. We should think that bad kings behave badly but that God represents the ideal king.

Whether people do or do not completely agree with Calvin's ideas of God as sovereign, the most appealing idea of God as king is to think that he has supreme power he uses for good. That good would be the good of God's subjects. We can interpret the biblical view to be that God is most like a good king—only better.

Husband

It is surprising, maybe startling, to see the title of husband for God. We find in Isaiah 54.5, "For your Maker is your husband, the LORD of hosts is his name."

How should we respond to the title of husband for God? Men especially are likely to be both uncomfortable and puzzled. How could God be a husband for men? There are problems also for women. Should married women think of God as a second husband? Is there a proper way in which unmarried women might think of God as husband? How appropriate is this title for God?

Jeremiah provided help for understanding the title of husband for God when Jeremiah wrote about a new covenant. There would be a difference from the old covenant, "a covenant that they broke, though I was their husband, says the LORD" (31.32). According to Jeremiah's understanding, the marital comparison does not picture God as a husband to individuals. The idea is that God is, figuratively, a husband for the whole Hebrew people. As a wife should be faithful to her husband, the Israelites should have been faithful to God.

What about the husband's faithfulness? Is all the responsibility supposed to be on the side of the wife? Hosea emphasized God's faithfulness as a spiritual husband to the people. In the Old Testament book bearing his name, Hosea had an unfaithful wife by the name of Gomer. Hosea literally redeemed Gomer, paying a price for her when she had fallen into bad circumstances. The prophet said in Hosea 3.2, "So I bought her for fifteen shekels of silver and a homer of barley and a measure of wine." Hosea was a faithful and loving husband in spite of Gomer's unfaithfulness. The larger point is that God is faithful even if his people are not. Hosea strikingly showed his faith that God is like a faithful husband and will redeem his people.

The title of husband for God at first does look strange, even unsettling. We see that the title has a powerful meaning. The people, like a wife faithful to her husband, are to be faithful to God. When the people are unfaithful, God is still faithful and, according to Hosea, can be forgiving.

Father

The title of father for God appears often in both the Old Testament and the New Testament.

In the Old Testament the writer of Psalm 68.5 spoke of God in this way: "Father of orphans and protector of widows is God in his holy habitation." We read in Psalm 103.13, "As a father has compassion for his children, so the LORD has compassion for those who fear him." We find more than one title in Isaiah 64.8: "Yet, O LORD, you are our father; we are the clay, and you are our potter; we are all the work of your hand." Jeremiah referred to God as saying, "for I have become a father to Israel, and Ephraim is my firstborn" (31.9). Malachi asked, "Have we not all one father?" (2.10). (Malachi may have been speaking of God, but possibly was referring to Abraham as father of the Hebrews.)

When we come to the New Testament, there are additional references to God as father, especially heavenly father. (See, for example, Matt. 5.16, 45, and 48.) Jesus prayed, "Our Father in heaven, hallowed be your name" (Matt. 6.9). (See also Matt. 7.21.) The parable of the prodigal son in Luke 15.11-32 could also be called the parable of the forgiving father. It is generally understood that the father in the parable represents God. Paul wrote in 2 Corinthians 1.3, "Blessed be the God and Father of our Lord Jesus Christ, the Father of mercies and the God of all consolation." In Ephesians 4.6, Paul referred to "one God and Father of all, who is above all and through all and in all." The writer of 1 John 1.3 stated, "See what love the Father has given us, that we should be called children of God; and that is what we are."

We may compare God as father, like God as king, to humans who have the same title. There are both good and bad human fathers, however. Many fathers provide safety, shelter, food, guidance, and emotional support. Others are harsh, neglectful, abusive, or absent. Some fathers are heroic while other fathers are terrible. Memories of good fathers are treasured. Memories of difficult fathers can be painful. Many fathers, being human, have some combination of both good and bad qualities.

With the varied experiences people have of human fathers, thinking of God as father can be complicated. Many people will bring their own experiences of human fathers to their consideration of God as father. Good associations with human fathers can stimulate positive thoughts of God, whereas bad associations with human fathers may interfere with trust in God.

The ideal is to think of God as a good father. God does not have any of the negative characteristics sometimes associated with human fathers. God represents the best of human fathers and exceeds even that comparison. The biblical writers viewed God as a good father in being responsible for human life and in caring deeply for his children.

Conclusion

Muslims traditionally have had ninety-nine names for Allah. Jews and Christians have no official number of titles for God. The main titles used by the biblical writers for God include creator, judge, redeemer, shepherd, king, husband, and father.

There are additional titles for God in the Bible for those who wish to consider them. The psalmist, for example, said in Psalm 18.2, "The LORD is my rock, my fortress, and my deliverer, my God, my rock in whom I take refuge, my shield, and the horn of my salvation, my stronghold." We have emphasized the titles that refer directly to persons.

The titles for God considered here have important meanings. Creator invites great respect for the power involved in bringing the world into existence. Judge implies accountability to God as the creator of humans. Redeemer provides hope for all who do not meet the standards required by God as the judge of the world. Shepherd carries clear indication of God's provision for God's people. The title of husband for God at first appears to be awkward and uncomfortable. There are, however, the strong meanings of desired faithfulness to God and also forgiveness by God for unfaithfulness. The titles of king and father can be complicated by the examples of people who have these titles. Believers trust that God does not have the bad qualities of human kings and fathers but has only the good qualities. God would be like a good king who uses his great power for the benefit of his people. God would be like a good father who lovingly provides for his family.

The Bible's titles for God add significant information to the Bible's defining qualities for God. The titles generally involve more of a direct comparison to humans than the defining qualities do. While some humans may dishonor their titles, the biblical view suggests that God is the ideal for the titles.

Questions for Further Thought

1. What additional interpretations do you have, if any, for one or more of the seven titles for God covered in this chapter?
2. How do you think your view of God may be influenced by the examples of people who have the same titles?
3. If some of the titles have greater meaning or appeal for you than others, why is that?
4. How would you explain the importance of the descriptive titles for God compared to the importance of the defining qualities?

CHAPTER 3

Divine Actions

In addition to defining qualities and descriptive titles for God, the biblical writers refer to divine actions. These reported actions by God provide further important information for thinking about God in relation to the Bible. The biblical accounts of God's actions strongly impressed G. Ernest Wright and Reginald H. Fuller. These men wrote their views in *The Book of the Acts of God*. With respect for that title, what do we find in the Bible about God's actions?

While there are many actions by God in the Bible, some actions have special significance. The Old Testament writers emphasized actions by "the God of Abraham, the God of Isaac, and the God of Jacob" (Exod. 3.15). These actions include creating, commanding, punishing, saving, making covenants, and calling. The New Testament writers stressed actions by "the God and Father of our Lord Jesus Christ" (Eph. 1.3). God's actions in the New Testament include incarnating, raising Jesus from the dead, and sending the Holy Spirit. Additional actions of God in the New Testament are revealing, reconciling, and working for good.

There have been different responses to these actions in different religions. Jews accept the actions of God in the Hebrew Scriptures but not those claimed in the New Testament—at least not those related to Jesus. Christians believe in God's actions throughout the Bible, both in the Old Testament and the New Testament.

Let us consider significant divine actions throughout the Bible.

The Beginning

The first two chapters of Genesis tell about God's action of creating. God created the heavens and the earth, light, vegetation, creatures of the sea and air, animals, and people. What was the value of what God had done? The writer of 1.31 said of the whole creation, "God saw everything that he had made, and indeed, it was very good."

Following God's creative activity was God's commanding activity. The first commandment was for others also to create in their own way. In the account of the creation of male and female persons, God said to them, "Be fruitful and multiply, and fill the earth and subdue it" (1.28). God gave an additional command for his creatures not to stray from what he had given them. In the story of Adam and Eve in the Garden of Eden, we read, "And the LORD God commanded the man, 'You may freely eat of every tree of the garden; but of the tree of the knowledge of good and evil you shall not eat, for in the day that you eat of it you shall die' " (2.16-17). The continuing story reveals that the man and the woman broke this additional command by eating of the tree of the knowledge of good and evil. (See 3.1-7.)

God's action of punishing came after the breaking of what God had commanded. Although the man and the woman did not immediately die physically, they died spiritually in their relationship to God. He cast them out of the Garden of Eden. (See 3.22-24.) The punishment included a great increase in woman's pain in childbearing. (See 3.16.) For the man, there was to be the curse of toil before he could eat from the ground. (See 3.17-19.) These verses could be interpreted with the view that the writer was trying to explain the origin of various human experiences. The direct meaning is that God punished those who disobeyed his commands.

God's actions do not end with punishing. There is also saving action by God. Although God cast Adam and Eve out of the Garden of Eden, he saved their physical lives at the time.

God's actions of punishing but also saving appear further in the story of Cain and Abel. God punished Cain for killing Abel but spared Cain's life. God's punishment for Cain was for Cain to be cursed from the ground. No longer would the ground yield its strength. The punishment sounds similar to the punishment for Adam. Also, Cain would be a wanderer on the earth. (See 4.11-12.) Yet there is saving action by God for Cain. God provided protection for Cain: "And the LORD put a mark on Cain, so that no one who came upon him would kill him" (4.15b). The mark on Cain, whatever it looked like, was not part of God's punishing activity. It was part of God's saving activity, that Cain would be saved from being killed.

Creating, commanding, punishing, and saving are all important actions of God in the beginning. The creating activity for the whole world occurred only at the beginning. God's actions of commanding, punishing, and saving continue to happen in other parts of the Bible.

Noah

The story of Noah shows God's commanding activity, punishing activity, and saving activity along with his covenant-making activity. The story appears in Genesis 6-8.

God decided he would punish the wickedness of men by blotting out the humans he had created. God would send a destructive flood, but Noah found favor with God. God commanded Noah, "Make yourself an ark of cypress wood; make rooms in the ark, and cover it inside and out with pitch" (6.14).

The covenant-making activity of God occurs in 6.18 when God addresses Noah: "But I will establish my covenant with you; and you shall come into the ark, you, your sons, your wife, and your sons' wives with you." A covenant is an agreement between parties. In the covenant with Noah and in other covenants, God takes the initiative and sets the conditions. God expects the other party to accept and to follow the covenant. The other party may break a covenant, but God does not. In some cases, God renews a covenant. In this case, Noah obeyed God and met the terms of the covenant.

Before the flood, Noah and his family and many animals went into the ark. When the flood came, God's punishment for a multitude of wicked people who were not in the ark was severe: They all drowned. Because Noah found favor with God, obeyed God's command to build an ark and followed God's covenant, God saved Noah and his family and the animals from destruction.

In this story, God commanded Noah to build an ark. Noah obeyed. God also made a covenant with Noah that required Noah and the members of his family to go into the ark. Noah again obeyed. When the ark was closed, God punished wicked people with a great flood. Although there was widespread destruction, God also had made provision for the saving of Noah, his family, and many animals. The story has God's commanding activity, covenant-making activity, punishing activity, and saving activity.

Other Early Stories

The remainder of Genesis has additional accounts of divine actions. The story of the tower of Babel in chapter 11 is another example of God's punishing activity. The problem was that some men tried to build a tower with its top in the heavens. God was not pleased with the project, possibly because the builders were trying to exceed their proper limitations. God punished the men by confusing their languages and by scattering them over the face of the earth. This story is another one that could be interpreted concerning origins, this time as an explanation for the diversity of languages. Whatever additional meanings there may be, God's punishing activity is prominent

A further example of God's covenant-making activity occurs in the story of Abram in Genesis 12. We also see calling activity and commanding activity. Although the word "covenant" is not specifically used, God called Abram by way of a command with an intended agreement. We find in verses 1-2: "Now the LORD said to Abram, 'Go from your country and your kindred and your father's house to the land that I will show you. I will make of you a great nation, and I will bless you, and make your name great, so that you will be a blessing." God called Abram and commanded him to go to a place where God wanted him to be. Abram's part of the agreement was to go. God's part of the agreement was to bless Abram.

The account of the destruction of Sodom and Gomorrah is a striking example of God's punishing activity along with his saving activity. (See Gen. 18.16-19.28.) God sent fire and brimstone to destroy the two wicked cities. Because of his mercy, God saved Lot and some of Lot's family members from the destruction. Lot's wife also would have been saved, but in the story she looked back when she was not supposed to and was punished by being turned into a pillar of salt

The story of Joseph in Genesis 37-50 is a profound illustration of God's saving activity. Joseph's jealous brothers sold him into slavery. Joseph arrived in Egypt and had some initial misfortunes. He later rose to become second only to Pharaoh, the ruler of Egypt. Joseph was able through seven years of plenty to prepare for seven years of famine. His brothers and his father, Jacob, eventually came to Egypt to escape the famine in Canaan. When Joseph's brothers learned that the one responsible for aiding them was Joseph, they were ashamed and frightened. Joseph told his brothers, "And now do not be distressed, or angry with yourselves, because you sold me here; for God sent me before you to preserve life" (45.5). God saved lives through Joseph. God saved the Egyptians of the time and Joseph's family from starvation.

Moses

Calling, commanding, punishing, saving, and covenant making all appear as God's actions in the story of Moses.

Exodus 3 has the story of how Moses saw a burning bush that was not consumed. God used the bush in his call to Moses: "When the LORD saw that he had turned aside to see, God called to him out of the bush, 'Moses, Moses!' And he said, 'Here I am'" (3.4).

What was the purpose of the call? God had a command for Moses. God said to Moses, "So come, I will send you to Pharaoh to bring my people, the Israelites, out of Egypt" (3.10). After many years in Egypt, the Israelites had become slaves and were treated harshly.

In Exodus 7-12, there is the account of punishment after Pharaoh would not let the people go despite the pleas of Moses. God punished the Egyptians with ten plagues, one after the other, the last being the death of many who were firstborn. The purpose of the punishment was to liberate the Israelites, that is, to save them from slavery.

Both punishing and saving by God appear again in the story of how the Israelites crossed the Red Sea. When Moses stretched out his hand over the sea, the waters parted. The Israelites crossed on dry ground. When the Egyptians tried to do the same thing, the waters returned and drowned them. By punishing the pursuing Egyptians, God had saved the Israelites from being recaptured and possibly killed. We read in Exodus14.20, "Thus the Lord saved Israel that day from the Egyptians; and the Israelites saw the Egyptians dead on the seashore." In 15.1-2 there is this report of a song sung by Moses and the Israelites to the LORD: "I will sing to the LORD, for he has triumphed gloriously; horse and rider he has thrown into the sea. The LORD is my strength and my might, and he has become my salvation."

God's covenant-making activity again occurs, this time with Moses. The writer of Exodus 19.5 said that Moses was to give the Israelites this message from God: "Now therefore, if you obey my voice and keep my covenant, you shall be my treasured possession out of all of the peoples."

The people were supposed to obey God's voice and keep his covenant by following God's commands. God's activity of commanding is powerfully expressed in the Ten Commandments. (See 20.1-17.) In addition, God commanded various ordinances. (See Exodus 21-23.) God further commanded specific instructions for building a tabernacle and for the practice of animal sacrifice. (See Exodus 25-40.) Moses was the leader both in receiving the commandments and in urging the people to follow them.

Prophets

Included in God's actions in the prophetic writings are God's punishing activity, God's saving activity, and God's covenant-making activity.

Isaiah understood some suffering by the Israelites as an indication of God's punishing activity. After the Israelites went into the promised land of Canaan, they had some success but eventually encountered devastating experiences. The united kingdom of the twelve Jewish tribes lasted only a short time historically and then divided. Both the northern kingdom of Israel and the southern kingdom of Judah came to brutal ends. The Assyrians destroyed the northern kingdom, while the Babylonians later conquered the southern kingdom. What had happened? Isaiah thought God was punishing the Israelites, at least the northern kingdom, for breaking their part of the covenant with God. The prophet wrote in Isaiah 10.5-6b: "Ah, Assyria, the rod of my anger—the club in their hands is my fury! Against a godless nation I send him, and against the people of my wrath I command him." Jeremiah 22.8-9 also implies the thought of divine punishment for abandonment of the covenant by the Jews.

Jeremiah looked forward to a time of God's saving activity and wrote of a new covenant God would make. This prophet proclaimed, "But this is the covenant that I will make with the house of Israel after those days, says the LORD: I will put my law within them, and I will write it on their hearts; and I will be their God, and they shall be my people" (31.33). The understanding is that God will save his people in spite of themselves. He will save them by helping them to do what they cannot or will not do on their own.

The Gospels

When we go from the Old Testament to the New Testament, there is a different emphasis concerning God's actions. The Old Testament writers saw God's actions primarily in relation to the earliest people and the Israelites. God's actions especially involved Adam and Eve, Noah, Abraham, Moses, the prophets, and the Jewish people. The Gentiles are not completely neglected, but neither are they given much attention. The New Testament writers thought of God's actions especially in relation to Jesus Christ. There were implications for all people.

Two New Testament gospels, Matthew and Luke, relate God's action in the conception of Jesus. It was through the action of the Holy Spirit that Mary, the mother of Jesus, conceived. Christians interpret these stories as referring to God's incarnating activity, his coming in human form.

Matthew's gospel has the account of an angel's appearing to Joseph in a dream and saying, "Joseph, son of David, do not be afraid to take Mary as your wife, for the child conceived in her is from the Holy Spirit. She will bear a son, and you are to name him Jesus, for he will save his people from their sins" (1.20-21). Matthew's story connects God's incarnating activity in Jesus with God's saving activity. Jesus, the incarnation of God, will save people from their sins.

Luke wrote of the angel Gabriel's speaking to Mary about God's incarnating activity: "The Holy Spirit will come upon you, and the power of the Most High will overshadow you; therefore the child to be born will be holy; he will be called Son of God" (1.35).

All four of the Gospels indicate God's action in sending the Holy Spirit to Jesus when Jesus was baptized. In Mark 1.10-11, for example, we read the following about Jesus, "And just as he was coming up out of the water, he saw the heavens torn apart and the Spirit descending like a dove on him. And a voice came from heaven, 'You are my Son, the Beloved; with you I am well pleased.'" (See also Matt. 3.13-17, Luke 3.21-22, and John 1.29-34.)

The Gospels have a strange omission concerning God's action. All four have accounts of the resurrection of Jesus after his crucifixion, but they do not tell us something we might expect. None of them clearly attributes the resurrection of Jesus to action by God. However, we do find such statements in other sections of the New Testament.

The Book of Acts

The book of Acts contains a reference to God's action in giving the Holy Spirit on the day of Pentecost. (See 2.1-4.) The recipients at this time were the disciples of Jesus. According to 2.4, "All of them were filled with the Holy Spirit and began to speak in other languages, as the Spirit gave them ability."

In his message that day, Peter interpreted the giving of the Holy Spirit at that time in relation to a prophecy from the book of Joel. The prophet spoke of God's future action with a very large group of people. Peter quoted from Joel 2.17b, "In the last days it will be, God declares, that I will pour out my Spirit upon all flesh." Perhaps the speaking in other tongues or languages on the day of Pentecost was an indication of the importance of the Spirit for increasing numbers of people.

Peter's message that day had a further important recognition. There was God's action in the resurrection of Jesus. Peter said about Jesus, "But God raised him up, having freed him from death, because it was impossible for him to be held in its power" (2.24).

The Letters of Paul

There have been many scholarly discussions on the authorship of letters attributed to Paul. Our concern here is to consider divine actions mentioned in the New Testament writings that are generally associated with Paul. There are several such actions, including raising Jesus from the dead, incarnating, reconciling, and working for good.

Paul clearly wrote of God's action in raising Jesus from the dead. There is the statement to the Thessalonian Christians that they had decided "to serve a living and true God, and to wait for his Son from heaven, whom he raised from the dead—Jesus, who rescues us from the wrath that is coming" (1 Thess. 1.9b-10).

The apostle connected God's incarnating activity with God's reconciling activity. Paul had this profound statement about Jesus: "For in him all the fullness of God was pleased to dwell, and through him God was pleased to reconcile to himself all things, whether on earth or in heaven, by making peace through the blood of his cross" (Col. 1.19-20). A brief statement by Paul is that "in Christ God was reconciling the world to himself" (2 Cor. 5.19).

Paul also mentioned God's working for good. According to some ancient texts, Paul wrote that "God makes all things work together for good" or that "in all things God works for good" (Rom. 8.28, alternate translations). A further indication of God's working for good is what Paul quoted in 1 Corinthians 2.9: "What no eye has seen, nor ear heard, nor the human heart conceived, what God has prepared for those who love him."

The Letter to the Hebrews

The letter to the Hebrews includes references to God's revealing action, creative activity, and action in the resurrection of Jesus.

The writer of the letter to the Hebrews speaks of God's revealing action in the prophets and in the Son. There is also a statement about God's creative activity through the Son. According to 1.1-2, "Long ago God spoke to our ancestors in many and various ways by the prophets, but in these last days he has spoken to us by a Son, whom he appointed heir of all things, through whom also he created the worlds." The sequence of events is profound. God's speaking through the prophets was followed by God's speaking through the Son, who preceded the prophets as an agent of creation.

The writer of the letter added his voice to other declarations that God acted in Jesus' resurrection. There is the reference to "the God of peace, who brought back from the dead our Lord Jesus" (13.20).

The writer of Hebrews sees God's actions in creating the worlds through the Son, speaking through the Son, and then raising the Son from death.

The Book of Revelation

When we come to the book of Revelation, we find a very complicated work with much symbolism that allows various interpretations. Without trying to understand everything or even much, we can find a significant example of God's divine action.

The writer of Revelation stated his faith in God's future action of providing good. The good will be in the form of comfort, as noted in 21.4a: A loud voice from the throne proclaimed that God "will wipe away every tear from their eyes." The Bible begins with the divine action of creating and has near the end the divine action of comforting.

Conclusion

God's actions in the Bible begin in Genesis with the account of God's bringing into existence a good creation. There are early statements and also further stories of such actions as commanding, punishing, and saving. We read further in the Old Testament about God's calling activity and the making of covenants.

The actions of God covered in the New Testament usually concern Jesus. These actions include incarnating, raising Jesus from the dead, and reconciling. There are also reports of sending the Holy Spirit and working for good. In the last book of the Bible, there is the hope for God's future action of providing comfort.

When we think about the actions of God, we can see that most of them are beneficial to humans. God created a good world and put humans in charge. God gave commands for human welfare. Failure by humans to follow the commands did bring punishment by God. Then there was action by God to bring remedies. In the Old Testament we find the making of covenants and the calling of great leaders. In the New Testament we witness the incarnation of Jesus, who was raised from the dead and brought reconciliation. Except for God's punishing activity, the biblical writers believe God's actions are generally intended for the happiness of people.

Questions for Further Thought

1. What other actions of God in the Bible should be considered in addition to the ones mentioned here?
2. How would you explain any differences between God's actions in the Old Testament and God's actions in the New Testament?
3. Why do you think that all of God's actions in the Bible should or should not be thought of as equal in importance?
4. How might thinking about God's actions in the Bible help us in thinking about God's actions beyond the Bible?

CHAPTER 4

God and Violence

While the Bible provides important beliefs about God in terms of defining qualities, descriptive titles, and divine actions, there are some areas of special concern. One special concern is the connection between God and violence in the Bible.

We usually think of violence as bad. It is force that almost always brings harm of some kind. We may have difficulty associating the harm of violence with God since God is said to be good. We even read that God is love. It may be troubling then to find many references to God and violence in the Bible. These references are in both the Old Testament and the New Testament, although the references are more numerous and stronger in the Old Testament.

How should we understand the biblical references to God and violence? A completely satisfactory explanation is difficult to find, but there are some possible answers. We can consider various views of violence. We can also think about biblical ideas of God's violence in relation to punishing actions, testing, and redemption.

Attitudes Toward Violence

Violence is widespread. Humans encounter violence in automobile accidents, crimes, and wars. Bad family relationships can lead to domestic violence involving arguments, fighting, and sometimes deaths. Animals encounter violence from other animals in the food chain. Some animals die, killed for food, so that other animals may live. The planet experiences violence in storms, floods, earthquakes, lightning strikes, and volcanic eruptions. Giant stars explode in outer space. Violence occurs throughout the universe.

Because violence usually brings damage of some kind, our attitude toward violence is ordinarily dislike and disapproval. We do not wish to see things, animals, and people hurt and sometimes destroyed. We especially do not like violence against ourselves and those who are close to us.

In spite of the harmful results of violence, we may sometimes approve of it. Violence can be used in some circumstances to achieve good as well as harm. How could violence be good? We may think of violence in a positive way if it is the best or perhaps only means for achieving a good goal. A dentist pulls a tooth to remove an infection and restore health to a mouth. Some harm occurs, but the final result is good. A surgeon cuts open a living body in order to correct a problem and perhaps save a life. In medicine there may sometimes be no other way to achieve a good goal than to use controlled violence. We can recognize that harm is done but be glad that good results outweigh the bad. With this view, you might choose to have your leg amputated rather than lose your life to gangrene. Beyond medicine, you might push or yank a person out of the path of a speeding car. The rescued person could be bruised and shaken but would be alive.

Although violence may be accepted or at least tolerated if it is used for a good end, we usually disapprove of it. We think violence is bad when there is a bad goal, such as to express undeserved hatred. There are usually objections to gratuitous violence, when there is no apparent reason for its use. We protest if violence is used even for a good goal when a nonviolent way is available. Why go through radiation or surgery if medication is available? Why spank a child if a reward system can bring good behavior? Violence is also objectionable if it is excessive, far more than is appropriate. Further, we do not approve violence that harms the innocent unless, perhaps, a vast amount of good can be achieved. Our attitudes toward violence can vary from sometimes acceptable to usually objectionable, according to the situation.

We are, however, not always consistent in our attitudes toward violence. Is the violence directed toward someone we like or someone we do not like? We may approve of violence against the bad guys but not like it against the good guys. Of course, not everyone agrees on who the good guys are and who the bad guys are. We usually find it hard to approve of any violence against ourselves or our loved ones. We may not be too troubled by violence that occurs to people far away, especially when we do not even know them. If some amount of violence is at least tolerable in some situations, we may not always be consistent in how much violence we may think is too much.

People can change in their attitudes toward violence. India provides an example of how at least some people may change attitudes toward violence. The concept of *ahimsa*, which means nonviolence or noninjury, goes back thousands of years in that country. Three of India's religions (Hinduism, Buddhism, and Jainism) have a strong heritage of nonviolence. Westerners are sometimes puzzled by the positive expression of nonviolence in respect for all life. Hindus, for example, have a high regard for cattle and usually let them roam without restriction. A Jain holy man will sometimes wear a cloth mask over his mouth. The goal is to avoid breathing in and injuring even tiny organisms in the air.

Mohandas Gandhi followed a great tradition when he rejected violence in his campaign to gain India's political independence from England. When the English used violence against Gandhi's followers, the English began to appear as bullies. They seemed to be mistreating peaceful people who simply wanted freedom for their own land. With nonviolence as a policy, the native people of India gained their independence from England in 1947.

When India gained its independence, it also went through partitioning. The process divided India with the goal of having majority Hindus and minority Muslims live peacefully in separate countries. Some of western India became West Pakistan. Some of eastern India became East Pakistan. West Pakistan later became Pakistan, and East Pakistan later became Bangladesh.

What happened when independent India was partitioned? Feelings of hostility erupted. In spite of India's ancient and noble teaching of ahimsa, many Hindus and Muslims killed each other during the period of relocation. Now mostly Hindu India and mostly Muslim Pakistan are rival nuclear powers. They continue to disagree, sometimes violently, over such issues as the control of Kashmir.

The example of India shows how attitudes toward violence can shift. Hindus had many centuries of teachings in favor of nonviolence. Yet many Hindus turned to violence against rival Muslims even when there were plans for peace. Muslims, who do not have a strong tradition of nonviolence, also attacked Hindus.

Christians are another example of changing and conflicting attitudes toward violence. Over the years there has been a great change from avoidance to widespread acceptance of violence.

The earliest Christians may or may not have been pacifists, but they usually did not wish to serve in the Roman army. Christians had the teaching of Jesus to turn the other cheek. In the first few centuries of the new religion, Christians were often the nonviolent victims of violent persecution.

A change in attitude may have started when Christians gained toleration in the fourth century. Then Christianity became the official religion of the Roman Empire. Christians started to approve military power and violence. Augustine of Hippo developed the "just war" doctrine. When Muslims spread across North Africa and into Spain, Christians became alarmed. Charles Martel, the grandfather of Charlemagne, stopped the Muslim advance into France. He led a Christian army to victory over the Muslims near Tours in 732. Christians and Muslims again fought one another during the Christian crusades of the early Middle Ages. The issue was control of the Holy Land, especially Jerusalem.

When Christians disagreed among themselves, they sometimes went to war against each other. It was mostly Catholic Christians and Protestant Christians who fought one another after the Protestant Reformation. Sometimes the differences were more political than religious, but each side could claim divine favor. In various wars when Christians from different countries fought one another, each group would claim that God was on its side.

Christians have disagreed over the violence of war itself. Many Christians have regarded violence in self-defense as a Christian duty. Would nonviolence have stopped Adolf Hitler in the terrible days of World War II? Yet Christian pacifists in World War I and World War II thought they were following the nonviolent teachings of Jesus by refusing to fight.

With the differing and often changing attitudes toward violence, we can recognize that violence is a difficult subject. There can be additional difficulty when we try to think about God and violence in the Bible.

Early in the Old Testament

An early reference in the Bible to God and violence occurs in the story of the flood in the days of Noah. (See Genesis 6-8.) Except for Noah and his family and animals that Noah took into the ark, people perished by drowning in the waters brought by God. We might see God's mercy in the sparing of some, but the destruction of others was great. Was no nonviolent way available? Did violence come to some innocent people? Was the violence excessive? We are not given answers to these questions.

The basic explanation for God's use of violence in the days of Noah is found in Genesis 6.11: "Now the earth was corrupt in God's sight, and the earth was filled with violence." Because of corruption and human violence, God decided to use violence to destroy the people. The belief is that God showed his moral goodness by using divine violence to punish human corruption. There was then an end to human violence, at least temporarily. But the people were not completely destroyed, and human violence later resumed.

Another early example of God's violence toward the wicked is the story of the destruction of Sodom and Gomorrah. (See Gen. 18.16-19.29.) At this time God did not direct violence toward all humans (except for Noah and his family), but toward the people of two cities. Would all in those cities be destroyed? Here we have a concern about the extent of violence. Surely God would not punish the innocent. Abraham asked, "Will you indeed sweep away the righteous with the wicked?" (Gen. 22.23). After much talk about numbers, God promised not to destroy Sodom if as few as ten righteous men were found there. We are not told anything about a final count, but the minimum number apparently was not met. Only Lot and his wife and his two daughters escaped when the LORD rained sulfur and fire from heaven. Lot's wife did not survive, because she looked back and became a pillar of salt. Many unrighteous people were killed, but God had allowed a possibility of escaping the violence if there had been enough who were righteous. Here is another account of belief in God's moral goodness through his violent punishing of the unrighteous, while mercifully allowing the possibility of escape.

A further major example of God's violence is the treatment of the Egyptians when the Israelites were seeking escape from slavery. There were the ten plagues, including the death of the firstborn. (See Exod. 7.14-12.32.) Why did God punish all of the Egyptians, including presumably innocent children, when it was Pharaoh who would not let the Israelites go? And why was it that Pharaoh sometimes hardened his heart but at other times God hardened Pharaoh's heart? (See, for example, Exod. 11.10.) Then, after the Israelites were allowed to leave, there was the drowning of the pursuing Egyptians at the Red Sea. (See Exod.14.10-29.)

We do not have all of our questions answered, but an explanation for this violence against the Egyptians appears in Exodus 14.31: "Israel saw the great work that the LORD did against the Egyptians. So the people feared the LORD and believed in the LORD and in his servant Moses." God's violence against the Egyptians not only secured the release of the Israelites from captivity but also strengthened the faith of the Israelites in God and Moses.

In these stories the main explanation for God's violence is punishment of the wicked. In the story of violence against the Egyptians there were also the good results of liberation and strengthening of faith for the Hebrews. Even if some innocent Egyptians were violently treated in the ten plagues, there was great good for the Israelites.

Animal Sacrifice

God's violence in the Old Testament is associated with animals as well as with humans. The books of Exodus and Leviticus contain God's commands for a system of animal sacrifice. There were violent and bloody killings of doves, sheep, goats, and bulls. Israelites provided for the sacrifice of animals in the tabernacle, a big tent that could be taken down and moved during their years in the wilderness. Then there was the sacrifice of animals in the successive temples in Jerusalem. The Jewish sacrifice of animals ended when the Romans destroyed the remaining Jewish temple in A.D. 70.

Why would God approve such violence against animals? There are at least two explanations. One is that God was honored when people gave something of value to him. The animals were supposed to be the best of their kind. It was not appropriate to sacrifice weak or sick or handicapped animals. An animal with a blemish would not have properly honored God. A second explanation is that the sacrifice of animals would be a penalty for sin. People would not suffer physically. The animals would bear the punishment. Individuals would pay a price economically by giving an animal or purchasing an animal to be sacrificed. According to this second explanation, violence ordered by God against animals allowed punishment for human sin by substituting animals for people.

The story of Abraham and Isaac in Genesis 22 allows an additional perspective concerning animal sacrifice. According to the Genesis account, Abraham was willing to sacrifice his son and was about to do so. God intervened at the last moment and provided a ram for the sacrifice instead. Should we interpret one meaning of the story to be that God wanted people to practice animal sacrifice rather than human sacrifice? Some of the early Israelites may have engaged in human sacrifice, especially the sacrifice of children. (See 2 Kgs. 23.10.) Although bad for the animals, the sacrifice of animals was better for humans than human sacrifice.

Whatever explanations may be found for God's approval of ritual violence toward animals, Jewish prophets spoke against animal sacrifice. Amos claimed to speak God's wishes, "Even though you offer me your burnt offerings and your grain offerings, I will not accept them; and the offerings of well-being of your fatted animals I will not look upon" (Amos 5.22). Another prophet said in Hosea 6.6, "For I desire steadfast love and not sacrifice, the knowledge of God rather than burnt offerings." Micah 6.8 asks, "He has told you, O mortal, what is good; and what does the LORD require of you but to do justice, and to love kindness, and to walk humbly with your God?"

God's reported ordering of violent animal sacrifice appears to have included a redemptive purpose in the forgiveness of human sins. And animal sacrifice was obviously better for humans than human sacrifice was. Yet the opposition of some Jewish prophets to animal sacrifice showed strong objections to the practice. The prophets thought there were better ways to honor God, ways that emphasized moral behavior and not violence.

The Promised Land

There were two prominent periods of violence after the Hebrews came to the promised land of Canaan: (1) when they were conquering and (2) when they were being conquered. According to the Bible, God was involved with the violence in both periods.

Joshua led the Hebrews in their invasion of Canaan. The Israelites believed God had promised the land to Abraham and his descendants. (See Gen. 12.1.) The Israelites thought God approved, even required, the violent conquest of the land. People in the land did not have the same understanding. They probably did not appreciate being invaded, dispossessed, and killed.

Why did God not lead the Israelites to a good but uninhabited land? Why did God lead the Hebrews to take land by force from people who already lived there?

According to Deuteronomy, the people of the land were wicked and deserved to be conquered and displaced: "When the LORD your God thrusts them out before you, do not say to yourself, 'It is because of my righteousness that the LORD has brought me in to occupy this land'; it is rather because of the wickedness of these nations that the LORD is dispossessing them before you" (9.4). In Deuteronomy 18.9-12 we read examples of this wickedness, including making a son or daughter pass through fire and also divination. And in Numbers 33.50-52, the Lord voices explicit objection to the practice of idolatry by the inhabitants of the land. The idea was not only to give the Hebrews some land but also to punish the earlier inhabitants.

As an example of God's direct participation in violence against enemies of the Hebrews, we may consider the aid given to the Gibeonites by Joshua against the Amorites. According to the report, the Israelites slaughtered many Amorites but not as many as God did. We read the following about the Amorites in Joshua 10.11: "As they fled before Israel, while they were going down the slope of Beth-horon, the LORD threw down huge stones from heaven on them as far as Azekah, and they died; there were more who died because of the hailstones than the Israelites killed with the sword."

As an example of God's commanded violence when the Israelites were in the promised land, consider the Amalekites. According to 1 Samuel 15.2-3, the LORD said to Saul through Samuel, "I will punish the Amalekites for what they did in opposing the Israelites when they came up out of Egypt. Now go and attack Amalek, and utterly destroy all that they have; do not spare them, but kill both man and woman, child and infant, ox and sheep, camel and donkey."

Saul did not follow the command completely. He spared Agag, the king of the Amalekites. Saul also kept back the best of the flocks, supposedly for sacrifice to God. The prophet Samuel thought there should have been complete destruction. Samuel said to Saul, "Has the LORD as great delight in burnt offerings and sacrifices, as in obedience to the voice of the LORD? Surely, to obey is better than sacrifice, and to heed than the fat of rams" (1 Sam. 15.22).

God's violence in the promised land came not only against the enemies of Israel but also against the Israelites themselves. Why would God punish his own people, the ones he had directed to the land they thought he had promised to them? Might it seem that God was interfering with his own plan?

The violence against God's people, after centuries in the promised land, was severe. There was a violent overthrow of the northern kingdom of Israel by the Assyrians, and there was a violent overthrow of the southern kingdom of Judah by the Babylonians. Why did these things happen? Jewish prophets thought God was ultimately responsible for the violence and was punishing his people

Isaiah wrote about God's violence concerning the northern kingdom of Israel, "Ah, Assyria, the rod of my anger—the club in their hands is my fury!" (Isa. 10.5) Jeremiah made this statement about God's violence concerning the southern kingdom of Judah: "Indeed, Jerusalem and Judah so angered the LORD that he expelled them from his presence" (Jer. 52.3). (See also Jer. 13.12-14 and 21.3-6.)

Why was God so upset with the people that he brought punishment upon them? Although the answer can be phrased in different ways, the main problem was that the people had broken their covenant with God. He had been faithful, but they had not. Jeremiah reported that God would make a new covenant that would be different from the one the people broke. Jeremiah wrote this message from God: "It will not be like the covenant that I made with their ancestors when I took them by the hand to bring them out of the land of Egypt—a covenant that they broke" (31.32)

In the promised land, God brought punishing violence against the enemies of Israel because the enemies were understood to be wicked. God later brought punishing violence against the people of Israel because the people were unfaithful to God and their covenant with him. The belief appears to be that God was good by giving the people the violent punishment they deserved.

The Book of Job

The book of Job contains a story about the violence God allowed toward a man from the land of Uz. This violence was not intended as a punishment for sin because Job was innocent. The violence came as a test of Job's faithfulness to God. Job passed the test. The author of Job disagreed with the widespread view that all violence against people was God's punishment for sin.

According to Job 1, the LORD asked Satan to consider God's servant Job, a blameless and upright man. Satan wondered how well Job would do if Job encountered misfortune. God allowed Satan to bring disasters upon Job. There was the killing of Job's animals and even Job's sons and daughters. But Job did not sin or blame God.

Then in Job 2, Satan presented another challenge: Let Job suffer physically as well as emotionally. The LORD then gave Satan further power over Job, except that Job's life was to be spared. Satan then afflicted Job with loathsome sores from the bottom of Job's foot to the top of his head. Job's wife wanted him to curse God and die, but Job still did not sin.

Much of the rest of the book of Job contains speeches by Job and friends who have come to see him. The friends are sympathetic, but they believe Job is suffering because of wrongdoing on his part. Their advice is basically for him to repent and seek God's forgiveness. Job does not think he has done anything wrong.

When we come to the end of the book, Job has an encounter with the LORD, who speaks to Job out of a whirlwind. (See 38.1.) We might expect that the LORD would give Job an explanation of what had happened, would tell Job that he had passed the test, and would congratulate Job. Instead God overwhelms Job and speaks of God's own mighty power. God asks where Job was when God did all of his mighty works. Job is intimidated. Perhaps Job thought that he had defended his genuine innocence too vigorously. Job said to God, "I know that you can do all things, and that no purpose of yours can be thwarted. . . . Therefore I have uttered what I did not understand, things too wonderful for me, which I did not know" (42.2-3).

Job did survive his ordeal and came out well. According to 41.1, "the LORD restored the fortunes of Job." There were more blessings than he had in the beginning. (See 42.12.) Job received money and gained more animals and more children.

What should we understand from this story of God and violence concerning Job? There is the view that God sometimes does not directly bring violence but allows it. The person who receives the violence may be innocent or, at least, not completely deserving of what happens. God may be testing the person but does not need to defend himself or give an explanation. It is good for the person who is a victim of violence to remain faithful, even if the person does not fully understand the situation. There is the strong belief that violence allowed by God may be testing rather than a punishment for sin.

The Crucifixion of Jesus

When we think about New Testament teachings, the emphasis is more on nonviolence than violence. Jesus taught that we should turn the other cheek and love our enemies. (See Matt. 5.39, 44.) Jesus spoke against violence when he said to Peter at the time of Jesus' arrest, "Put your sword back into its place; for all who take the sword will perish by the sword" (Matt. 26.52). Jesus advocated mercy rather than violent revenge for those who put him to death. According to at least some ancient texts, Jesus said, "Father, forgive them; for they do not know what they are doing" (Luke 23.34). Paul and Peter both advocated not returning evil for evil. (See Rom.12.17 and 1 Pet. 3.9).

In spite of these teachings about nonviolence, the New Testament also has references that connect God with the violent death of Jesus by crucifixion. How could it be that God was involved with such violence?

If we think of who was responsible for wanting Jesus to be crucified, we go back to some Jewish and Roman leaders. According to the Gospels, some Jewish leaders thought Jesus was guilty of blasphemy and therefore deserved death. Since Jews under Roman rule could not carry out the death penalty, Jesus was taken to the Roman governor, Pontius Pilate. Although Pilate was not concerned about disputes among the Jews, Pilate was ready to deal with any real or perceived threat to Roman rule. Because some thought of Jesus as King of the Jews, Jesus was a possible threat to the establishment. Pilate ordered the crucifixion of Jesus. Roman soldiers carried out the order.

How could anyone think God was involved in the violence of the crucifixion? As strange as it may seem, Jesus himself is reported to have had such a thought. When Jesus went out to the Mount of Olives before he was betrayed, he prayed, "Father, if you are willing, remove this cup from me; yet, not my will but yours be done" (Luke 22.42). The cup had various associations. In this case, the cup was something Jesus preferred to avoid: his coming suffering. Since Jesus prayed nonetheless for God's will to be done, the indication is that Jesus thought it was God's will for Jesus to suffer through crucifixion.

Also, when Jesus appeared before Pilate, the Roman governor asked if Jesus did not know that Pilate had power to crucify Jesus and power to release Jesus. According to John 19.11, Jesus replied, "You would have no power over me unless it had been given you from above." Can there be any other meaning for Jesus' statement but the idea that God gave Pilate power that included the power to crucify Jesus?

Further, at the time Peter was addressing Jews from every nation on the day of Pentecost, he said of Jesus that "this man, handed over to you according to the definite plan and foreknowledge of God, you crucified and killed by the hands of those outside the law" (Acts 2.23). While the actual crucifixion was done by "those outside the law," Peter thought that Jesus was handed over for crucifixion "according to the definite plan and foreknowledge of God."

Why might God have planned the violent crucifixion of Jesus? There are many theories. The overall idea is that Jesus was the innocent Son of God but died a redemptive death for others.

One view of Jesus' crucifixion is that God allowed Jesus to be the perfect sacrifice for our sin. There was animal sacrifice by the Jews even during the lifetime of Jesus, but the writer of the letter to the Hebrews saw a problem. He wrote in Hebrews 10.4, "For it is impossible for the blood of bulls and goats to take away sins." He added about Jesus, "But as it is, he has appeared once for all at the end of the age to remove sin by the sacrifice of himself" (Heb. 9.26). This writer believed that the one-time sacrifice of Jesus was far superior to animal sacrifice.

There are additional views. There is the idea that Jesus died as a substitute for us. He bore the suffering we deserved. Paul said in 1 Corinthians 15.3 "that Christ died for our sins in accordance with the scriptures." There is the interpretation that Jesus' death on the cross was the supreme example of God's love for us. Paul wrote in Romans 5.8, "But God proves his love for us in that while we still were sinners Christ died for us." Other explanations for Jesus' death include a ransom (see Mark 10.45 and 1 Tim. 2.6), a satisfaction for God's offended honor (Anselm), reconciliation (see Rom. 5.10 and Col. 1.19-20), and victory over sin, death, and the devil (see Heb. 2.14-15). (For a more detailed account of the meaning of Jesus' death, see my book *What Should We Believe About Jesus?*)

Jesus and others did have teachings in the New Testament against violence between humans. In addition, there are strong indications that God planned the violent death of Jesus to bring redemption.

Conclusion

The likelihood that violence will result in harm usually leads us to a negative view of violence. We especially object to violence when it is undeserved and excessive. Yet there can be acceptable uses of violence when the result is more good than harm and when there is no other way to achieve a needed goal. What do we find about God and violence in the Bible?

There is a close relationship between God and violence in the stories and teachings of the Old Testament. Also, there is usually some explanation for the violence. The main explanation is that God used violence to punish the wicked, including his own people when they were unfaithful. We read that God sent a flood in the days of Noah because the people were wicked. God sent fire and brimstone to the evil cities of Sodom and Gomorrah. He punished ancient Egyptians with plagues and drowning for making slaves of the Hebrews and not setting them free. He ordered a violent system of animal sacrifice so that animals could be punished instead of people. When the Israelites came to the promised land, God punished their enemies for wickedness and idolatry. God later punished his own people for their unfaithfulness to their covenant with him.

In addition to punishment of the bad, God's violence in the Old Testament sometimes had beneficial results. The violence against the Egyptians secured the release of the Israelites from captivity and strengthened the Israelites' belief in God.

In the book of Job, God allowed Satan to bring violence against Job not for punishment but for testing. Job was innocent and remained faithful in spite of his suffering.

Several references in the New Testament indicate God willed and planned the violent crucifixion of Jesus. Although we usually object strongly to the use of extreme violence against an innocent person, New Testament writers believed that the death of Jesus brought enormous spiritual benefit. God's arrangement for this violence was intended to bring great good..

The relationship between God and violence in the Bible continues to be challenging, but there are some reasonable explanations.

Questions for Further Thought

1. What do you believe about the idea of some acceptable uses of violence?
2. What do you see as the differences, if any, between Old Testament statements about God and violence and New Testament statements about God and violence?
3. How would you explain the relationship between biblical statements about God's violence and statements about God's love and mercy?
4. In what ways do you think biblical views about God and violence may apply to people today?

CHAPTER 5

God and Human Destiny

In addition to God and violence in this life, a second special area of concern for thinking about God in relation to the Bible is human destiny. What does God have planned for the future of humans beyond this life? The question is important both for our view of God and our interest in what will happen to us. One frequent answer is that God has planned heaven beyond this life as a wonderful experience for many. Another common answer is that God has also planned hell, often thought of in very violent ways, as punishment in the next life for those who reject him in this life. There are additional views.

For purposes of comparison, we will first consider some major beliefs outside of the Bible about human destiny. These views include the immortality of the soul and reincarnation.

When we turn to the Bible, we find several ideas about human destiny. The Old Testament has references to Sheol, heaven, and the resurrection of the dead. There are also statements suggestive of hell. The New Testament has teachings about the resurrection of the dead, heaven, and hell. There are various references to the soul throughout the Bible.

Christians have had different interpretations of the meaning of these biblical teachings. There have also been some variations from traditional views, at least partly due to different understandings about God.

Immortality of the Soul

Belief in the soul and its immortality is an old belief. For example, the Greek philosophers Socrates (469-399 B.C.) and Plato (428-348 B.C.) emphasized the immortality of the soul. Plato expressed their views in his dialogue "Phaedo" with Socrates as the main speaker.

There was little doubt for Socrates that the soul existed. The question was how long it would exist. Would it be possible, for example, for the soul to outlast a body, perhaps even several bodies, but then fade away? Socrates was convinced of the immortality of the soul and had several arguments for that view. Although the various arguments are probably interesting only to philosophers, Socrates's main argument for the immortality of the soul was that the essence or basic nature of the soul is life. Socrates reasoned that the soul could not change its basic nature and become the opposite essence, which would be death. Therefore, Socrates reasoned, the soul would always exist.

Socrates thought that the soul's immortality called for great care to be given to the soul. He did indicate some hesitation regarding his arguments, suggesting less than absolute certainty. But, he said, the risk of believing is noble.

If the soul is considered to be life, it is easy to understand the view of Aristotle, the main pupil of Plato. Aristotle believed that every living thing had to have a soul in order to live. He thought that even animals and plants, as living things, had to have souls.

Unlike Aristotle, Socrates spoke only of humans, not of animals and plants. He questioned what happens to the human soul. What is its destiny? Socrates thought that death for people was the separation of the soul from the body. Whether or not the separation was good depended on the person. Socrates mentioned two possibilities.

A philosopher, a lover of wisdom, would regard the separation of the soul from the body as good. He would not be concerned about the body and its pleasures, but he would be interested in the soul and its wisdom. For Socrates, truth was not perceived through the bodily senses but through the reason of the soul. Socrates associated both life and thought with the soul. The philosopher would rather be free of the body. For his part, Socrates hoped his soul would go where he could talk to gods and good men. That situation would have been heaven for Socrates. He did not explain how a soul could talk without a body.

In contrast, a philosoma, a lover of the body, frets when he is about to die. Separation of his soul from a body would be bad for him. If his present body is worn out, he would like for his soul to come back into another body. He would like to be reincarnated.

Epicurus (341?-270 B.C.), another Greek philosopher, emphasized a life of calm and rational pleasure but did not believe in existence after death. As a materialist, he did not believe in the immortality of the soul or even in the soul itself.

Christians have had mixed views about the immortality of the soul. There are important biblical references that seem to support the idea of the soul. There is, for example, the story of the creation of man in Genesis 2. God formed man from the dust of the ground, but man did not become a living being until God "breathed into his nostrils the breath of life" (v. 7). According to tradition, "breath of life" is interpreted as the soul and the soul is immortal.

Yet many Christians, as well as some Jews, have emphasized the resurrection of the body. There is not the view that a dead body simply comes to life again but that the earthly body is transformed after death. Concentration on the immortality of the soul often involves low regard for the body and hope of separation from it, similar to what Socrates taught. Belief in the resurrection of the body frequently includes conviction of the perfection of the body.

While belief in the immortality of the soul and belief in the resurrection of the body are different, there does not appear to be a strong reason as to why they could not be combined. Some may wish to accept both beliefs.

Reincarnation

Many religious teachers of the Far East have emphasized reincarnation. Hinduism, Buddhism, and Sikhism all include the belief that the soul goes from one body to another body at death. This process of transmigration of the soul from body to body may occur, according to some interpretations, hundreds or even thousands of times. Karma determines what the next reincarnation will be, whether better or worse. Karma has several meanings, but basically it refers to the law that actions have consequences without exception. The sum of a person's actions in a body becomes that person's karma, which will lead to either a better body or a worse one in the next reincarnation.

Although people with these beliefs seek improvement from one body to the next one, the ultimate goal is to seek release from reincarnation. Each bodily existence involves suffering, so the goal is to break the chain or wheel of existence. A person ends suffering by not being born again.

Nirvana is release from reincarnation and the suffering involved with it. Nirvana is usually considered indescribable. There is great happiness, but also the individual ceases to exist. It is like blowing out the flame of a candle or having a drop of water absorbed into the ocean.

How does one arrive at nirvana? Some Hindus and Buddhists believe that higher beings can assist individuals toward nirvana. Some Hindus think that devotion to one's deity, a particular god or goddess, has great value. Followers of Pure Land Buddhism seek gracious help from Buddha Amitabha or Amida. The stricter view in Hinduism and Buddhism is that nirvana comes only after an individual goes through intense meditation over a long time. Sikhs, most of whom live in India, believe that nirvana occurs by means of a person's love for the one God.

In contrast, biblical Christians believe in being born again but do not believe in reincarnation. They are talking about a spiritual birth that occurs once rather than a physical birth. Christians believe that a person has only one lifetime in one body. It is the spiritual birth that allows entrance into heaven.

Sheol in the Old Testament

When we turn to the Old Testament, the most frequent view of human destiny is Sheol. The word is usually a synonym for death. It can also mean a gloomy place for the spirits or shades of all of the dead. The concept of Sheol implies immortality of the soul but conveys the idea of a shadowy, sad existence after death.

In Genesis 37.35, when Jacob thought that his son Joseph was dead, Joseph refused to be comforted and said, "No, I shall go down to Sheol to my son, mourning." Job expressed the finality of Sheol when he said in Job 7.9-10, "As the cloud fades and vanishes, so those who go down to Sheol do not come up; they return no more to their houses, nor do their places know them any more." In Psalm 49.13-14, we read of "the fate of the foolhardy ... Like sheep they are appointed for Sheol; Death shall be their shepherd; straight to the grave they descend, and their form shall waste away; Sheol shall be their home." In referring to the arrogant, the writer of Habakkuk 2.5b stated, "They open their throats wide as Sheol; like Death they never have enough."

The Old Testament references to Sheol are numerous. Most are similar to the ones already mentioned. (See 2 Sam. 22.5-6; Job 11.8, 26.6; Pss. 9.17, 18.5, 55.15, 139.8; Prov. 5.5, 7.27, 9.18, 15.24, 23.14; and Isa. 14.9, 28.15, and 18.)

Although the references to Sheol are usually negative, the psalmist expressed an optimistic thought in Psalm 49.15: "But God will ransom my soul from the power of Sheol, for he will receive me." (See also Ps. 16.10.) The writer of Isaiah 25.8a had a similar thought when he said of the LORD that "he will swallow up death forever." It would be good if we had more information about what these writers were thinking.

The writer of Ecclesiastes expressed doubt about what would happen after death, apparently including Sheol. He asked, "Who knows whether the human spirit goes upward and the spirit of animals goes downward to the earth?" (Eccl. 3.21).

In the Septuagint, the Greek version of the Hebrew Old Testament, the Hebrew word *Sheol* is translated as the Greek *Hades*. In ancient Greek mythology, Hades was the god of the underworld. His name also designated his kingdom, the realm of the dead. Sheol and Hades are similar in referring to the place for departed souls, but the words are not identical in what they cover.

According to ancient stories, the Greek gods provided special punishments for a few in Hades. Those who greatly displeased the Greek gods and had unusual suffering included Tantalus, Sisyphus, and Prometheus. Tantalus was hungry and thirsty and could see food and drink, but they were always beyond his reach. Sisyphus had to roll a large rock or small boulder up a hill. The rock would then roll down the hill, and Sisyphus had to go through the same experience over and over again without end. Prometheus, having angered the gods by giving fire to humans, was chained to a rock and had his liver eaten by a bird of prey. There was a new liver each night, so Prometheus had the same experience daily. The Old Testament differs. It does not have any stories about special punishment by God for those in Sheol.

If we had only the Old Testament references to Sheol, we could get the idea of the immortality of the soul or spirits or shades. All of the souls would have basically the same fate, which would be a gloomy, sad existence. The Old Testament, however, includes additional beliefs.

Heaven in the Old Testament

Old Testament references to heaven are comparatively few and usually indicate the sky or the place where God is. We read in Genesis 1.1 that "God created the heavens and the earth." It is possible to interpret this verse as referring to more than one heaven, but the usual interpretation of "heavens" is the area above the land. In Exodus 16.4, the LORD spoke to Moses about raining "bread from heaven." The meaning appears to be that bread will come from above or the place where God is. Then there is the story of how Elijah called for fire to come down from heaven and consume some soldiers. (See 2 Kgs. 1.9-12.) The fire supposedly came from God.

The Old Testament does have some striking suggestions of heaven as a place of destiny for certain individuals. Although there is no direct mention of heaven, the writer of Genesis 5.24 said, "Enoch walked with God; then he was no more, because God took him." Did God only take him from life or take him to heaven? The reference in 2 Kings 2.11 is more explicit where we read that "Elijah ascended by a whirlwind into heaven."

In spite of some intriguing possibilities concerning two individuals, the Old Testament does not contain explicit teachings about heaven as a place of happiness prepared by God for those who have died in his favor.

Bodily Resurrection in the Old Testament

There are various references in the Old Testament that may be understood in relation to the resurrection of the body. Some passages are more definite than others.

It is not clear how much support is provided by Ezekiel 37 for a general belief in the resurrection of the body. We find in that chapter the story of the dry bones that came together and received sinews, skin and breath. We read in verse 11 of God's saying that "I am going to open your graves, and bring you up from your graves, O my people; and I will bring you back to the land of Israel." Is there more here than imagery of the return of Jews from exile in Babylon? We do read about the return of some Jews to Jerusalem, but there is no account in the Old Testament that dead Jews in Babylon were literally raised from their graves to live again.

There is this powerful statement in Job 19.25-26: "For I know that my Redeemer lives, and that at the last he will stand upon the earth; and after my skin has been thus destroyed, then in my flesh I shall see God." It is tempting to think of this verse as referring to the resurrection of the body. Was Job perhaps thinking instead of how his body was afflicted by loathsome sores (the destruction of his skin) and how he was trusting that his body (his renewed and healthy flesh) would still survive to see God?

Although some passages are subject to differing interpretations, the Old Testament has other sections with clearer indications of belief in the resurrection of the body. Consider this strong statement in Isaiah 26.19: "Your dead shall live, their corpses shall rise. O dwellers in the dust, awake and sing for joy!" The writer of Daniel 12.2 gave a prophecy of resurrection along with indication of different destinies: "Many of those who sleep in the dust of the earth shall awake, some to everlasting life, and some to shame and everlasting contempt."

Jews have disagreed over belief in the resurrection of the dead. The Pharisees and the Sadducees differed over this point as well as some other issues. According to Acts 23.8, "The Sadducees say that there is no resurrection, or angel, or spirit; but the Pharisees acknowledge all three." (See also Matt. 22.23.) The Sadducees accepted as authoritative only the books of Moses, the first five books of the Old Testament. Those books do not mention the resurrection of the body.

The Old Testament has beliefs that do not seem to go together, at least not easily. There is belief in Sheol, which implies immortality of the soul without a body. There is also belief in the resurrection of the body. We are not told exactly how the two beliefs may be combined.

Hell in the Old Testament

The Old Testament has at least the beginnings of belief in hell as a place of severe punishment for the wicked. Hell is not only different from heaven but also different from Sheol. As we have seen, Sheol designates either the condition of death or the place of the dead or both. Sheol contained the shades or spirits of all of the dead, both good and bad, without physical bodies. Sheol was a place of gloomy darkness. Suffering there was not due so much to imposed pain but to lack of the blessings of earthly existence. Hell differs from Sheol in containing only the wicked and in imposing much more severe suffering.

We do not find a word for hell as such in the Old Testament. We find instead references to the valley of Hinnom. The Hebrew *gehinnom* is similar to the Greek *Geenna* or *Gehenna*, the main Greek word for hell.

What was written about the valley of Hinnom? In referring to reforms under King Josiah, the writer of 2 Kings 23.10 said, "He defiled Topheth, which is in the valley of Ben-hinnom, so that no one would make a son or a daughter pass through the fire as an offering to Molech." The valley was a location for burning children as a sacrifice. Josiah attempted to end the horrible practice. In addition, a prophecy of Jeremiah 7.32 stated how the valley of Hinnom would be used as a burial ground: "Therefore, the days are surely coming, says the LORD, when it will no more be called Topheth, or the valley of the son of Hinnom, but the valley of Slaughter: for they will bury in Topheth until there is no more room." The valley of Hinnom, located near Jerusalem, became a garbage dump where fires burned constantly.

The Old Testament references to the valley of Hinnom (or the son of Hinnom) do not provide a fully developed concept of hell. The references do associate death, burial, and fire.

Bodily Resurrection in the New Testament

We find more about the resurrection of the body in the New Testament than in the Old Testament. There are various kinds of accounts.

There are accounts of how Jesus raised people from the dead to renewed earthly life. Examples include the daughter of a ruler (Matt. 9.18-26), the son of a widow of Nain (Luke 7.11-17) and Lazarus, the brother of Mary and Martha (John 11.1-44). Presumably all of those people died again at some later time.

All four of the Gospels end with accounts of the resurrection of Jesus on the third day after his crucifixion. He appeared and disappeared at various places. We also read in Acts 1.1-11 that Jesus presented himself alive for forty days and then ascended into heaven. The indications are that his body was both resurrected and transformed. The disciples saw Jesus and talked to him, but he was not subject to earthly restrictions.

Paul had special comments about the resurrection of the body. The apostle had believed in the resurrection of the dead as a Pharisee (see Acts 23.6), and continued with that belief as a Christian. He tried to explain both the coming resurrection and the transformation of our bodies.

Paul's famous teaching on the resurrection of the body is found in 1 Corinthians 15. In verses 42-50, he distinguished between kinds of bodies. What is sown or dies as a physical body is raised as a spiritual body. Paul said there are different kinds of flesh or bodies for men, animals, birds, and fish. Why could there not be yet another kind of body? Paul's idea was that a human body can die but can be transformed into a spiritual body, something more than simply coming to life again. When and how the transformation takes place are mysteries. The idea of a celestial or spiritual body is a difficult concept and sounds inconsistent. If a body is material but a spirit is not material, how could there be a spiritual body? But Paul was trying hard to explain what he sincerely believed.

Paul connected the resurrection of others with the resurrection of Jesus. Paul declared in Philippians 3.20-21, "But our citizenship is in heaven, and it is from there that we are expecting a Savior, the Lord Jesus Christ. He will transform the body of our humiliation that it may be conformed to the body of his glory, by the power that also enables him to make all things subject to himself." Paul believed that the followers of Jesus would experience something similar to what happened with the resurrected Jesus.

Heaven in the New Testament

The New Testament contains various references to heaven as a place or condition beyond this life. There are both direct and indirect references.

Among the explicit mentions of heaven in the New Testament is this statement of Jesus in Matthew 5.11-12a: "Blessed are you when people revile you and persecute you and utter all kinds of evil against you falsely on my account. Rejoice and be glad, for your reward is great in heaven." As we have seen, Paul said in Philippians 3.20 that "our citizenship is in heaven."

We might wonder how many heavens or perhaps levels of heaven there are supposed to be. Paul had this curious statement in 2 Corinthians 12.2: "I know a person in Christ who fourteen years ago was caught up to the third heaven—whether in the body or out of the body I do not know; God knows." The writer of Revelation distinguished between an old heaven and a new heaven. We read in 21.1, "Then I saw a new heaven and a new earth; for the first heaven and the first earth had passed away."

Some references in the New Testament are associated with heaven, even when the word is not mentioned. There is this statement in Luke 16.22 about a poor man named Lazarus: "The poor man died and was carried away to be with Abraham." The Greek text reads "to Abraham's bosom." The assumption is that Abraham was in heaven. Additional references that indicate heaven without using the word include the promise of Paradise to the thief on the cross (Luke 23.43), eternal life (John 3.16), and "what God has prepared for those who love him" (1 Cor. 2.9).

The rich descriptions given in Revelation 21 are usually considered to be accounts of heaven but, strictly speaking, refer to "the holy city Jerusalem coming down out of heaven from God" (v. 10). It is the holy city, not heaven itself, that is pure gold with gates of pearls. (See vv. 9-21.)

The many references to heaven in the New Testament are, considered together, not easy to interpret. For example, how should we understand the third heaven, the new heaven, the new earth, and the holy city? The references do indicate something far better than earthly life.

Hell in the New Testament

The New Testament has various references to hell with use of the word "Gehenna." Consider some statements attributed to Jesus in the first Gospel. The writer of Matthew 5.22 reported a warning from Jesus that "if you say, 'You fool,' you will be liable to the hell of fire." We read these words in 10.28: "Do not fear those who kill the body but cannot kill the soul; rather fear him who can destroy both soul and body in hell." In 18.9, the Gospel writer quoted Jesus as saying, "And if your eye causes you to stumble, tear it out and throw it away; it is better for you to enter life with one eye than to have two eyes and to be thrown into the hell of fire."

In the famous story of the separation of the sheep and the goats in Matthew 25.31-46, there is no reference to Gehenna, but there is mention of "the eternal fire" (v. 41) The goats would be assigned to it. These references to the hell of fire and to eternal fire are consistent with the fires that burned in the valley of Hinnom when at least some of it became a garbage dump.

After reading about the hell of fire and eternal fire, we may be puzzled to find indications of darkness in the world to come. Fire and darkness do not seem to fit together. There are the references in Matthew's Gospel to "outer darkness." (See 8.12, 22.13, 25.30.) The writer of 2 Peter 2.4 said of sinful angels that God "cast them into hell and committed them to chains of deepest darkness." The Greek word for hell in this verse is *Tartarus* (the deepest level of hell) rather than Gehenna, but the idea of darkness is evident. Are we to think of hell as having a limited amount of fire with increasing darkness away from the fire? Should we think of hell as having at least two levels, one with fire and one with darkness?

In Matthew 16.18, there is the claim by Jesus that "the gates of Hades will not prevail" against his church. The traditional interpretation is that the forces of hell will not destroy Christ's church. But the reference is to the gates of Hades, not the gates of Gehenna. An important meaning of Hades is the realm of the dead, all of the dead. Also, gates are not weapons used for offense but are defensive structures intended for protection from outside forces. Gates are also designed to keep something inside. In view of these thoughts, a possible interpretation of Matthew 16.18 is that Christ's church will smash the gates of Hades and rescue the dead. The idea may be stated in an odd way but is intriguing.

In the famous story of Lazarus and the rich man in Luke 16.19-31, there is a reference to Hades rather than Gehenna. The account, however, is that the rich man died and was buried and then was in torment in Hades. Although one meaning of Hades is the realm of all of the dead in a shadowy existence, this story suggests a place of suffering for the wicked. Hades appears here to have the same meaning as Gehenna.

The book of Revelation does not contain the word "hell" (Gehenna) but does have forecasts of doom. There is the declaration that various beings will have the fate of the lake of fire. These beings include the devil, the beast, and the false prophet. (See 20.10.) Condemned also are all whose names were not written in the book of life. (See 20.15.) Specifically mentioned for "the lake that burns with fire and sulfur" are the cowardly, the faithless, the polluted, murderers, fornicators, sorcerers, idolaters, and all liars. (See 21.8.)

In addition, we find a puzzling declaration in Revelation 20.14: "Then Death and Hades were thrown into the lake of fire. This is the second death, the lake of fire." Since Hades is often a synonym for death, what does it mean to refer to both Death and Hades? Also, what is meant for Death and Hades to be thrown into the lake of fire? Could it be the death of death? If the lake of fire is hell, what is the meaning of throwing Hades into hell? The meaning of this verse is very difficult to understand.

The references to hell in the New Testament are sometimes direct and sometimes indirect. The teachings are strong but complicated. It is difficult to arrive at a completely consistent overall interpretation.

The Soul

There are various references to the soul throughout the Bible. We read, for example, about loving God with all your soul (Deut. 6.5 and Mark 12.30), lifting up the soul to the Lord (Ps. 25.1), and a rejection of offering the fruit of the body for the sin of the soul (Mic. 6.7). We also read about finding rest for your soul (Jer. 6.16 and Matt. 11.29), fearing the one who can destroy both soul and body (Matt. 10.28), and saving a sinner's soul from death (Jas. 5.20).

Unlike the case with Socrates and Plato, we do not find any specific teaching in the Bible about the nature of the soul. There is no section of the Bible on the immortality of the soul similar to Paul's teachings in 1 Corinthians 15 on the resurrection of the body. References in the Old Testament to Sheol imply belief in the immortality of the soul without directly stating the view. Also, the Bible does not have teachings that specifically connect the soul to God's plans for human destiny.

God's Involvement

The Bible has much to consider concerning human destiny, about what will happen to people after death. The teachings are numerous, but it is not easy to see how all of the teachings fit together. There is also comparatively little said directly in the Bible about God's involvement. Most of the references to human destiny in the Bible do not specifically mention God.

Because there is much in the Bible about human destiny but little directly said about God's involvement, it is not surprising that Christians have had different interpretations. Let us consider some of the varying views on God's plan for people to avoid hell and reach heaven.

Strict Calvinists have the most severe view. Calvinists believe that all humans deserve condemnation to hell but that God has mercifully spared some, electing or choosing a certain number for heaven. All others are destined for eternal punishment in hell. God shows great mercy by sparing anyone. Calvinists believe in a limited atonement, that Christ died only for those God chose. In the Calvinist view, the elect have nothing to do with their salvation; saving faith is not their own doing but a gift from God. Calvinists believe their interpretation properly gives all glory to God. For some others, the Calvinist view makes God look unfair by not giving the non-elect a chance to escape hell.

Evangelical Christians believe that God gives people a choice of their own destiny. They believe that God will admit to heaven all those who accept Jesus Christ as Lord and Savior, that it is God's wish for all to be saved and for no one to perish but that God has allowed people to make their own decision.

The traditional Roman Catholic view of human destiny has God in relation not only to heaven and hell but also purgatory. The basic conviction is that God has mercifully provided sacraments as the means of salvation through Jesus Christ. Traditional Catholics have emphasized the importance of dying in a state of grace. A baptized Christian dies in a state of grace if there are no unforgiven mortal sins, those considered grave or deadly. A dying Christian with a mortal sin, such as murder or adultery, may receive forgiveness through proper confession. A baptized Christian who dies with unforgiven venial sin, which is not grave or deadly, is not excluded from heaven.

According to traditional Catholic belief, those who die in a state of grace still have to suffer for their sins in purgatory. Only after being cleansed of their sins through suffering will they enter heaven. The only ones to avoid purgatory and go directly to heaven are martyrs and saints. Martyrs are those who have died for the faith. Canonized saints are those who receive official church recognition after performing at least two miracles.

For traditional Catholics, those who are not baptized go to hell at death. Those who have received Christian baptism but do not die in a state of grace also go to hell when they die.

Christians continue to be divided over how they think God deals with people concerning human destiny. It is likely that these differences will not be overcome anytime soon.

God's Justice and Love Regarding Hell

When we read about God in the Bible, some of God's qualities and titles have a close association with human destiny. One of the titles for God is judge, and we are supposed to think of God as a just judge. One of the defining qualities of God is love. There are conflicting opinions on whether or not God's justice and love are compatible with the traditional doctrine of hell.

According to one view of God's justice and hell, people deserve severe punishment by offending him with their sin. In other words, God is entirely just in giving them what they are due. How severe should the punishment be? Since God is infinite, there is the belief that God is infinitely offended by people's sin and should give them infinite punishment in hell. That punishment would be agonizing and eternal.

This idea of God includes a comparison of him to an absolute monarch. The classical explanation of this interpretation is Anselm's *Cur Deus Homo?* (*Why Did God Become Man?*). Anselm considered the feudal concept that an offense should be judged in relation to the rank of the person offended; specifically, a person of great rank would be more highly offended than a person of low rank—even for the same offense.

Ripley's Believe It or Not Museum in Gatlinburg, Tennessee, provides a historical example. There is the story that Catherine the Great of Russia ordered a severe punishment of twenty years for one of her noblemen. The jailers put the prisoner into a small cage in a kneeling position with his head tilted forward but without room for him to sit down, lie down, or stand. What had he done? His trousers had split when he was bowing to the queen. We might consider the happening unfortunate, even somewhat humorous. But he had offended someone of the highest rank in the land. It was the rank of the person offended that determined the severity of the punishment.

Would not hell be just punishment for everyone who has offended God, the highest of all? Although the Bible does refer to God's vengeance, this more severe view is not completely stated. According to Numbers 31.3, Moses told the people "to execute the LORD's vengeance on Midian." (See also Num. 32.35, 41, 43 and Rom. 12.19.) There is this plea in Psalm 94.1: "O LORD, you God of vengeance, you God of vengeance, shine forth!" (God is addressed twice.) We read in Nahum 1.2, "A jealous and avenging God is the LORD, the LORD is avenging and wrathful; the LORD takes vengeance on his adversaries and rages against his enemies." But none of these references claims that God imposes infinite or eternal punishment.

In a different view of God's justice and hell, we might expect God to follow the proportional understanding of justice. What would that understanding be? The great principle of proportional justice is the right amount—not too little and not too much. This principle appears in the Old Testament command of an eye for an eye and a life for a life. (See Exod. 21.23-24.) Equal punishment, according to this interpretation, is proper. Excessive or unlimited vengeance would not be proper. Might then agonizing and eternal punishment in hell be too much punishment? A possibility here is that severe punishment in hell might be appropriate but not excruciating punishment without end. How could a finite person commit more than finite sin even against God as infinite? Would not finite punishment be appropriate?

A problem for this view is why an admittedly good principle for humans should be applied to God. The principle of an equal amount, not more or less, sounds like a fair principle of justice for relationships between humans. Should the same principle apply to relationships between God and humans? Should Almighty God be restricted by any human standards?

Then there is the biblical teaching that God is love. Is the traditional view of hell compatible with this belief?

Many people are convinced that the traditional view of hell does not conflict with a God of love. Calvinists believe everyone would go to hell if it were not for God's love. Only the elect go to heaven, but God shows love for them. Evangelicals believe that God loves everyone and wants each person to go to heaven, and lovingly provided the way through Jesus Christ. It is sad when people do not accept God's love and do not love God in return. Catholics likewise think God loves all and wishes for all to love him. But traditional Catholics, like evangelicals, do not think God's love prevents severe punishment for those who reject him.

A differing opinion is that a loving God would not send anyone to hell and maybe would not have provided hell in the first place. If we thought that someone had devised a place for eternal torture, would we not consider that person barbaric? Surely God is not like a barbarian.

The difficulty felt by some over a loving God and hell can be expressed in the following imagined story. Suppose a father with several children decides to add two rooms to his house. One room has toys and ice cream. The other room has spiders, snakes, and scorpions. The father then goes away but leaves instructions for the children about the two rooms. The children become confused because people who talk to the children disagree on how to interpret the instructions. The children then learn that some of them will go to one room and some to the other room, permanently. Among other questions, might the children ask what kind of father provided the bad room? Would the children be puzzled if they were told their father was very loving? It is difficult for some to think that God would be like the father in the story.

As we have seen, there are strongly different opinions on the relationship between God's justice and hell as well as the relationship between God's love and hell. We might wonder if there can be a completely acceptable resolution.

Questions about God and Hell

There are many questions about God and hell. We will not consider all of them, but let us consider a few.

Did God plan Sheol and then change to hell (Gehenna)? In the Old Testament the most common reference to human destiny is Sheol. The place or condition designated by Sheol was for the shades or spirits or souls of all of the dead. They were in a gloomy and deprived but not otherwise painful state. Then we read in the New Testament about Gehenna. It contains only the wicked dead or those who do not meet God's requirements for entry into heaven. There is the pain of eternal agony. How should we understand the relationship between Sheol and Gehenna in God's plan?

What is the origin of hell? We read in the first verses of the Bible that God created the heavens and the earth. The Bible does not have a story about God's creation of hell. Perhaps the closest we come is the statement attributed to Jesus about "the eternal fire prepared for the devil and his angels" (Matt. 25.41). But the Bible does not have a clear explanation of the beginning of hell as a place of punishment for people.

Should we interpret the biblical references to hell literally or symbolically? If the references are to be taken literally, do they always refer to the next life or at least sometimes to the valley of Hinnom? If the references are symbolic, what is being symbolized?

What is the exact relationship between God and hell? It is difficult to find teachings in the Bible that directly link God and hell. Much of the doctrine about God and hell comes from interpretation, not direct biblical statements.

Adjustments Regarding Hell

With concerns about God's justice and love regarding hell and with some important questions about hell, should there be any adjustments in the traditional doctrine of hell? Some say no, and others say yes. Calvinists, for example, think they have the correct beliefs concerning God and hell. They do not see a need to change. Various others have adjusted their thinking about hell. The adjustments, explicitly or not, have been in line with biblical beliefs about God's love and mercy.

Augustine of Hippo in the fifth century emphasized the necessity of baptism for salvation. He reluctantly assigned dead but unbaptized infants to hell. Thomas Aquinas in the thirteenth century was a prominent advocate for the doctrine of limbo as an adjustment regarding hell. Aquinas agreed that dead but unbaptized infants would go to hell but believed they would go to limbo, a border or edge of hell. They would never experience the beatific vision, that is, never see God in heaven. But they would never experience any pain of sense, that is, any physical pain. Thomas Aquinas found a way to express God's mercy without rejecting an important doctrine.

Most Baptists are among those who think that lack of baptism will not send dead infants to hell. The general belief is that all infants who die will be allowed by God's mercy to go directly to heaven. While believing that followers of Jesus should be baptized, Baptists do not believe in the necessity of baptism for salvation. In the case of infants and young children, there is also appeal to "the age of accountability." The reference is not to an exact year but to the time when children are mature enough to tell right from wrong. Although the phrase is not in the Bible, the idea is expressed in Isaiah 7.15 as "the time he knows how to refuse the evil and choose the good." The belief is that God lovingly and mercifully does not hold humans responsible until they reach the age of accountability.

In 2007 the Vatican's International Theological Commission expressed reservations about the concept of limbo as being too restrictive. While limbo has been believed by many Catholics for many centuries, it was never official church dogma. The commission, after receiving approval from Pope Benedict XVI, said there was no sure knowledge but nevertheless prayerful hope that infants who died unbaptized might go to heaven.[1] For those who believe baptism is necessary for salvation, how could infants who die without baptism go to heaven? A possible explanation may be to consider baptism as usually necessary but not as absolutely necessary for salvation. Infants who die without baptism might be thought of as an exception allowed by God's mercy.

When we think of hell itself, might it be in accord with God's mercy to think of the fire of hell as symbolic rather than literal? In one interpretation of this view, hell may be eternal but does not have any physical suffering. There would be the terrible fate for all in hell of endless separation from God but no other pain. This idea extends the conditions of limbo to all of hell. But there is an additional consideration. William Crockett thinks the fire of hell is symbolic rather than literal but does not represent mercy. He does not know the exact nature of the punishment but believes it will be very bad.[2]

There is some support for a symbolic view of hell with the belief that the soul is a spiritual rather than material reality. If the soul is an immortal spiritual reality, how could it suffer any kind of physical punishment? But the soul supposedly could suffer the pain of eternal separation from God. A limitation for this symbolic view is an additional belief, the conviction that souls will eventually be reunited with bodies at the time of general resurrection and judgment.

Some faith groups have adjusted their thinking on the amount of time spent in hell. Universalists, for example, began by denying that punishment in hell would be eternal. They thought that those in hell would at some time have their punishment end. Eventually everyone would be saved and go to heaven. This view is somewhat like the doctrine of purgatory. There would be suffering for a time, but the suffering would not last forever. This idea can be traced at least as far back as the early church father Origen. It has been strongly suggested by the modern writer Rob Bell in *Love Wins*, in which he considers the claim that God wants everyone to be saved. Will not God finally get what God wants?

There is also the idea that God will, perhaps mercifully, annihilate the wicked. They would die but not suffer in hell. Seventh-Day Adventists and Jehovah's Witnesses believe in the destruction of the wicked. In addition, for example, Clark Pinnock thinks of hell in terms of annihilation. He believes that immortality is conditional, intended only for believers and not for those who reject God.[3] The idea of annihilation is similar to the belief in nirvana in that at least some individuals—the wicked in this case—would cease to exist.

But is there anything in the Bible about annihilation? Consider the plea in Psalm 104.35a: "Let sinners be consumed from the earth, and let the wicked be no more." Is the request that the wicked not only perish from this life but also have no life after death? There is this intriguing statement attributed to Jesus in Matthew 10.28: "Do not fear those who kill the body but cannot kill the soul; rather fear him who can destroy both soul and body in hell." Who but God could destroy both soul and body? Might God mercifully destroy both the bodies and souls of wicked people?

Will there be additional proposed adjustments in thinking about God and hell? It is hard to predict what those, if any, may be and whether or not they will gain much acceptance. Many, of course, continue to believe that there should not be any adjustments from their traditional views.[4]

Conclusion

As we began consideration of God and human destiny in relation to the Bible, we included some views outside of the Bible for comparison. We looked at belief in the immortality of the soul in Greek philosophy. We examined thoughts concerning reincarnation in Eastern religions. These views provide a background for examining biblical teachings about God and human destiny.

We have surveyed teachings about human destiny in the Bible. The Old Testament has statements about Sheol (the realm of all dead spirits), heaven (usually the sky or a place where God is), and the resurrection of the body. Old Testament references to hell do not fully develop the concept of a place or condition of severe punishment. The New Testament has statements about the resurrection of the body, heaven (as a wonderful place or condition), and Gehenna or hell (a place of suffering for those who do not meet God's requirements to enter heaven). There are various references in the Bible to the soul, but the references do not deal directly with immortality.

It is difficult to derive a completely clear and consistent doctrine of God and human destiny from the Bible as a whole. There are many statements about human destiny, but the connection between them and the connection between God and them is not always clear. Sheol in the Old Testament and Gehenna in the New Testament are not the same. The relationship between the immortality of the soul and the resurrection of the body is not fully explained. There is not a complete account of how God is involved.

Christians have been divided over how they think about how God has planned for people to avoid hell and gain heaven. Not many people are troubled about heaven, but there are many concerns about hell. There have been different opinions about God's justice and love in relation to hell and also questions about biblical teachings concerning hell itself. There have been adjustments by some Christians in their conceptions of hell.

Biblical teachings about God and human destiny are more complicated than we might first expect. With the many and varied teachings, it is understandable that there are questions and differing views. In spite of varying interpretations, Christians do believe that God is both just and loving and has made appropriate plans for human destiny whether we completely understand those plans or not.

Questions for Further Thought

1. What similarities and differences do you see between human destiny in the Old Testament and human destiny in the New Testament?
2. What do you believe is the proper interpretation of God's justice and God's love in relation to hell?
3. What is your opinion of the various adjustments some people have made regarding their view of hell?
4. How would you explain the main biblical teachings about God and human destiny to someone who is not a Christian?

Notes

[1] See http://www.catholicnews.com/data/stories/cns/0702216.htm.
[2] See William Crockett's essay "Metaphorical" in *Four Views on Hell*, ed. William Crockett (Grand Rapids: Zondervan, 1996).
[3] See Clark Pinnock's essay "Conditional" in *Four Views on Hell*.
[4] See, for example, John Walvoord's essay "Literal" in *Four Views on Hell*.

PART 2

***

God and Philosophy

CHAPTER 6

Trying to Prove God's Existence

When we turn from God and the Bible to God and philosophy, we encounter a different way of thinking. Philosophers usually reject an authoritarian approach and wish to have a rational basis for beliefs. Although reason can be and has been used with the Bible, philosophers generally prefer to use reason apart from the Bible when thinking about God.

If philosophers do not go by the Bible when they think about God, what do they use for their thinking? One possibility is the world of nature. There is some support in the Bible itself for thinking about God in this way. Even before there was a Bible, the psalmist declared in Psalm 19.1, "The heavens are telling the glory of God; and the firmament proclaims his handiwork." Paul wrote, "For what can be known about God is plain to them, because God has shown it to them. Ever since the creation of the world his eternal power and divine nature, invisible though they are, have been understood and seen through the things he has made" (Rom. 1.19-20).

In spite of the convictions of biblical writers, philosophers have been divided over whether God's existence can be proved by reasoning about nature or anything else. Some philosophers have offered what they have considered to be proofs for God's existence. These philosophers have included Anselm, Thomas Aquinas, and William Paley. Other philosophers and philosophical writers, even though they personally believed in God, have not thought that reason could prove God's existence. Included in this group are Blaise Pascal, Søren Kierkegaard, and Leo Tolstoy. Then there have been philosophers who have thought that God's existence could not be proved because, in their view, God does not exist. Albert Camus is one example.[1]

Although philosophers seek understanding, many philosophers themselves are often difficult to understand—and some philosophers are more difficult to understand than others. Their views on trying to prove God's existence show some of the difficulties.

Using a Definition

When Anselm (1033-1109) tried to prove God's existence, he did not reason from the world of nature but from a definition. Anselm was a native of Italy who became Archbishop of Canterbury in England in 1093. He presented his views in *Proslogium*. Both his ideas and his style of writing require careful attention.

Anselm started his proof for God's existence by defining God as a certain kind of being. He defined God as that being than which nothing greater can be conceived. Since ontology is the study of being, philosophers have called Anselm's views the ontological argument for God's existence.

There is the question, Anselm thought, of whether such a being exists. The fool in the Bible said there is no God. But even the fool believed the idea of God existed in the understanding. The fool did not believe God existed outside of the understanding. Anselm disagreed with the fool in the Bible.

Anselm gave two versions of his argument. His first version claimed that God must exist in reality and not only in the understanding. A being that existed in reality would be greater than a being that existed only in the understanding. If God existed only in the understanding, he would not be the greatest conceivable being. Anything that did exist in reality would be greater than God. But Anselm defined God as that being than which nothing greater can be conceived. Therefore, Anselm concluded, God must exist in reality and not only in the understanding.

According to the claim in the second version of Anselm's argument, it is inconceivable for God not to exist. We can conceive of a being that *can* be conceived not to exist. We can also conceive of a being that *cannot* be conceived not to exist. A being that cannot be conceived not to exist would be greater than a being that can be conceived not to exist. What kind of being is God? With Anselm's view that God is that being than which nothing greater can be conceived, we would have to think of God as that being that cannot be conceived not to exist. In Anselm's thought, God was the only being that cannot be conceived not to exist.

What should be said about the fool in the Bible, the one who said there is no God? Anselm explained that the fool apparently knew only the word that signified God. The fool did not know the meaning of the word or, at least, Anselm's definition of God. Anselm thought that anyone who understood the meaning of the word would believe that God exists.

Gaunilo was a French monk who disagreed with Anselm's argument. He claimed he was writing in defense of the fool. Although not challenging

God's existence, Gaunilo thought that Anselm's argument was ridiculous. As an example, Gaunilo suggested that we could conceive of an excellent island, even the greatest conceivable island. But conceiving of the island, however excellent or great, would not guarantee the island's existence. How could a conception or definition of something, no matter how great, prove its existence?

In his reply, Anselm agreed that Gaunilo had a good point. We do not usually think that something conceivable has to exist. But, Anselm thought, God is a special case. God is not like the greatest conceivable island. Even with the greatest conceivable island, we could still conceive something even greater. But God is that being than which absolutely nothing greater can be conceived. Other things can be conceived without having to exist, but Anselm believed that the greatest conceivable being had to exist. Anselm's view was that, even if he did not believe in God, his own argument would convince him that God exists.

How strong is Anselm's argument? A difficulty with Anselm's views is that he does not deal with the standards usually employed in determining existence. We ordinarily wish to see or touch or hear something before thinking it exists. While we can consider the idea of a being than which nothing greater can be conceived, Anselm gave no image or picture or description of God. He gave no clear content for us to consider, nothing for us to check by seeing or touching or hearing. Anselm had an impressive idea. But we usually do not think that the existence of an idea guarantees the existence of the object of the idea, especially an idea with little content.

Using the World

Thomas Aquinas (1225-1274) did not base his arguments for God's existence on a definition, as Anselm did. Aquinas considered what the apostle Paul considered, which was experience of the world. The ancient Greek philosopher Aristotle also greatly influenced Aquinas. The medieval theologian gave a Christian interpretation of Aristotle's ideas concerning experience of the world. Aquinas gave his arguments for God's existence in *Summa Theologica*.

Aquinas realized there were objections to the existence of God. For example, (1) evil in the world is incompatible with the infinite goodness of God, and (2) there is no need to suppose God's existence. All natural things could be explained by the principle of nature. All voluntary things could be explained by human reason or will.

Although Aquinas stated these objections, he did not agree with them. He agreed instead with the witness of the Bible that God exists. He proceeded to offer five proofs for God's existence in support of the biblical view.

Aquinas' first argument for God's existence is from motion. All of the arguments are "from" something about our experience of the world, which we can see, touch, and hear. Among other considerations, we can sense motion in the world. Aquinas observed that things do not move themselves but must be moved by something else. A movement of something has been brought about by a previous mover, which was brought about by a mover previous to it. How far back do we go in a succession of movers? Aquinas did not believe in an infinite regress of movers, a going backward forever without any beginning. He concluded there must be a first mover, moved by no other. That first mover is God.

The second argument is from causation and is similar to the first argument. We experience efficient causes in the world. However, a thing cannot be the efficient cause of itself but must be caused by something else. What caused something was itself caused. Each cause was brought about by a previous cause. Aquinas again rejected an infinite regress, this time of efficient causes. He concluded that there was a first efficient cause, which itself was not caused. That first cause would be God. It was God who started everything, including the world. Since cosmology is the study of the world, philosophers sometimes call this second argument the cosmological argument for God's existence.

The third argument is from possibility and necessity. Nature has things that are possible to be and possible not to be. If it is possible for everything not to be, then at some time there was nothing in existence. If at one time there was nothing in existence, nothing could begin to be and nothing would be in existence now. Since the world does exist, it is absurd to think that all things are possible to be and possible not to be. Aquinas concluded there must be something with necessary existence, or something that has to be. What necessarily existed allowed the existence of what is only possible to be. The necessary being, the one who has to exist, is God. It was God as necessary being that allowed the world to come into existence. Since this argument refers to how the world came to exist, philosophers have sometimes called the third argument the cosmological argument for God's existence. It can be confusing, but both the second argument and the third argument have been called the cosmological argument.

The fourth argument is from the gradation or degree that we experience in the world. There are various levels of more or less of something.

Aquinas did not supply many examples, but let us think of how there are different levels of strength. There are some beings with much strength and some with little strength. Aquinas reasoned that gradation or degree of more or less implies a maximum, that is, a highest degree. He believed the maximum would be the cause of the various grades and degrees. Aquinas thought the maximum of all being and goodness, which was also the cause of them, was God.

The fifth argument is from the governance of the world. There are things in the world that lack knowledge but that nevertheless act for an end or goal. They serve a purpose. Aquinas thought there was design when things lack knowledge but still act for an end. The existence of design implies a designer, a being with knowledge and intelligence. All of those things in nature that act for an end or have a goal must have had a great designer. God is that designer. Since teleology is the study of ends or goals, philosophers have usually called this fifth argument the teleological argument or argument from design.

Aquinas then replied to the original objections to God's existence. One objection was that God does not exist because the existence of evil is not compatible with the goodness of God. Aquinas replied that God and evil both exist but that God can bring good out of evil. The other objection was that God does not exist because God is not needed to explain nature. Aquinas answered that nature works for an end or goal and does need an explanation. That explanation is a higher agent or principle, namely, God.

The arguments of Aquinas have the advantage of starting with the world, with things that can be seen or touched or heard. Aquinas then reasons to something behind the world, something that cannot literally be seen or touched or heard. We can, for example, see movement but cannot literally see God as the First Mover. We can see beings that come into existence and leave existence, but we cannot literally see God as the only being that has necessary existence. The question then is whether we can reason from what is seen to what is unseen. Aquinas thought we could.

David Hume (1711-1776), a Scottish philosopher, saw a difficulty with the arguments of Aquinas. Hume gave his views in *An Enquiry Concerning Human Understanding*. Hume was an empiricist, one who emphasizes knowledge through sense experience. Since Aquinas also emphasized knowledge through experience, what was the problem?

Hume had reservations about causation, and all of Aquinas' arguments involve causation. The First Mover caused motion to begin. The First Cause started a series of causes and effects, including the world. The being who is necessary caused the existence of what is possible to be and possible not to be. The maximum of something causes various levels of something. A great designer causes things without knowledge to serve a purpose. Hume's problem with causation is that it is something we cannot see or touch or hear. How do we know that causation exists? For Hume, we do not know causation as such but have experience of regular and ongoing associations. Hume's skeptical interpretation of causation raised doubts for many philosophers about Aquinas' arguments.

Another difficulty with the arguments of Aquinas is their view of God. How could there be any difficulty? Do not the arguments all conceive of God in magnificent ways?

Think of titles for God based on the arguments. We would have First Mover, First Cause, Necessary Being, Maximum of Being and Goodness, and Great Designer. Those titles are all very impressive, but they are more abstract than personal. Does this God, the God of philosophy, sound like the God of the Bible? We read in the Bible about the God of Abraham, Isaac, and Jacob. We read about "the God and Father of our Lord Jesus Christ." Thomas Aquinas' conception of God in the arguments, while profound, does not have the historical and personal qualities of God found in the Bible. Defenders of Aquinas will say that it is the same God, but the perspectives are very different.

Using a Watch

William Paley (1743-1805), an English churchman who provided a version of the teleological argument or argument from design, compared the world to a watch and compared God to a watchmaker. According to his views espoused in *Natural Theology*, someone who found a watch in a place such as a meadow would need an explanation, a special explanation. The person could not explain a watch the same way that a stone might be explained. A stone could have been around forever. The stone would have no apparent purpose, even though it could be used in various ways. A watch, however, would have the end or purpose of telling the time. The only explanation would be that the watch had a maker who designed it for that purpose.

Paley thought the works of nature are like the watch in design. If anything is different, the works of nature show design to an even greater degree. There must have been a designer and maker for the watch. So there must have been a designer and maker for the world. As a watchmaker is the explanation for a watch, so God is the explanation for the world.

Paley's argument is not as complicated as some of the other arguments for the existence of God. Its comparative simplicity is appealing. But how strong is the argument?

Paley's kind of argument is called the argument from analogy, namely, that two things alike in one or more ways are probably alike in one or more additional ways. Paley claimed that a watch and the world both show evidence of design or purpose, a watch's design or purpose requires a watchmaker and, therefore, the world's design or purpose also requires a designer and maker.

But how similar are a watch and the world? It is evident that a watch has been designed for the purpose of telling the time. If the world has been designed for a purpose, what is that purpose? Such a purpose should not be assumed but clearly stated. To say that things in the world look designed because they have a purpose is not the same as saying that the world itself has a purpose. Paley's argument would have been stronger if he had claimed and explained a purpose for the world, a purpose that required a designer.

Believing with Reason

There are people who have not been persuaded by the various arguments for God's existence but have still believed in God. One such person was Blaise Pascal (1623-1662), a French mathematician and philosopher who expressed his views in *Pensées* (Thoughts). He thought that reason could not decide the existence of God, but claimed it was more reasonable to believe than to disbelieve.

Considered the founder of modern probability theory, Pascal associated belief in God with gambling, which apparently was of only mathematical interest to him. Anthologists sometimes include his views under the title of "The Wager."

Pascal said that reason cannot tell us if God does or does not exist. He did not challenge the power of reason in other areas but thought it was inadequate concerning God. We still must choose what to believe. It is somewhat like choosing heads or tails on the flip of a coin. Pascal was not saying that flipping a coin and deciding what to believe about God have the same importance, but both cases involve a choice. Why should we choose one way or the other? On the matter of God's existence, reason itself cannot decide which choice to make. But Pascal thought there was reason to consider belief in God as the better choice.

According to Pascal, our chances of winning or losing concerning belief in God were even. Reason does not compel us one way or the other.

We have just as much chance to choose correctly as to choose incorrectly. The odds, we might say, are 50-50. What we should do, according to Pascal, is consider the possible gain or loss. If we choose to believe in God and are correct, we will gain everything. We will have all the blessings God has prepared for those who believe in him. If we choose to believe in God and are incorrect, we will have lost nothing. We would be mistaken but would keep everything we had. Who would not take an even bet with everything to gain and nothing to lose?

For any who still had difficulty believing, Pascal advised them to act as if they believed. He was not trying to encourage hypocrisy but thought that acting as if one believed would help a person to believe. Pascal further thought that acting as if one believed would bring benefits in the present life. The person would try to live in accord with good qualities such as humility, gratitude, and generosity. The person would avoid the poisonous pleasures of glory and luxury. Pascal further supported his views by saying the heart has reasons that reason does not know.

Pascal's views will appeal to some but not to others. For those who are undecided, Pascal has an intriguing claim about the possibility of gaining everything and losing nothing. For those who think of gambling as evil, it may be troubling to think of a possible connection between God and any kind of gambling. For yet others, faith should be simple and childlike with no special attractions toward belief.

Believing with No Reason

Søren Kierkegaard (1813-1855), a Danish thinker, also did not believe that reason could prove God's existence. He believed in God for apparently no reason. In his *Philosophical Fragments*, Kierkegaard was not against reason but considered it insufficient to deal with God. Kierkegaard said that reason seeks to know God but cannot, because God is absolutely different from what reason can conceive. Reason brings God as near as possible, but God is still as far away as ever.

Kierkegaard claimed to reason from existence rather than toward existence. He would ordinarily not try to prove the existence of something. He would take something thought of as existing and try to prove its nature. Consider a court of law. The procedure is not to prove that the accused exists but to try to prove that the accused is guilty. The procedure is reasoning from existence to essence rather than reasoning from essence to existence. Kierkegaard did not mention Anselm but appears to have had Anselm's reasoning in mind. Anselm tried to reason from a definition of God, God's essence, to God's existence. Kierkegaard rejected that kind of reasoning.

For Kierkegaard, reasoning about God's existence does not work. Kierkegaard advocated a leap, sometimes referred to as a leap of faith. Letting go of the proof is a leap he understood as allowing the existence of God to be recognized.

Kierkegaard's views will appeal to those who wish to emphasize that God is beyond human comprehension and that God must simply be accepted. Kierkegaard's ideas will not appeal to those who wish to have a reason for what they do. How can there be belief in God without some kind of reason? Why should there be a leap? How does letting go of attempted proof allow God's existence to be recognized? Further, if God is completely beyond human reason, how could God be thought of at all? Kierkegaard's views lead to some difficult questions.

Believing Against Reason

The Russian writer Leo Tolstoy (1828-1910) was famous as a novelist, especially as the author of *War and Peace*. He expressed both personal and philosophical thoughts about God in his book *My Confession*, in which he made a confession of his faith. Like Pascal and Kierkegaard, Tolstoy did not find reason to be adequate in determining God's existence. He even thought that believing in God went against reason. Tolstoy did, however, find great help from belief in God.

Tolstoy had advocated living so as to derive the greatest comfort for oneself and one's family. Many share that goal, so what was the problem? He began to be troubled by thoughts of death and the meaning of life. He did have a good and loving wife and family. He was respected by friends and neighbors and was praised even by strangers. He had considerable physical and mental powers. (He claimed at one time that he could work like a peasant.) Yet he realized that sooner or later death would come to his family and to himself. Life began to appear to him as a stupid, mean trick played on him by someone. Tolstoy became very despondent and thought of suicide.

He sought answers to the meaning of life in science and philosophy, but was disappointed in both. While science provided answers to questions he had not asked, how would it help him, for example, to know the chemical composition of the stars? When it came to the meaning of life, science regarded life as a temporal, accidental combination of particles. Philosophy, at least the philosophical views Tolstoy studied, regarded the world and human life as incomprehensible.

Tolstoy decided that the meaning of life could be found only in the religious faith of simple people. He compared the beliefs and lives of the poor people to the members of his own circle. People in his group had great intelligence and material privileges but did not understand the meaning of life. Tolstoy's peers had much leisure but were bored and complained and had difficult deaths. Hardly one in a thousand was a believer. The poor people worked hard but found some amount of satisfaction and peace. They had a hopeful outlook even at the time of death. Hardly one in a thousand was not a believer. It was the religious faith of the poor people that made the difference.

Tolstoy thought that faith involved the negation of reason. But he believed that faith with the negation of reason was far better than the negation of life. Faith made living possible. Faith gave meaning to life by relating the finite to the infinite. Faith had answers. There is the question of how one should live. Faith answers that a person should live in accord with God's law. There is the question of the real result from one's life. Faith answers with the prospect of eternal torment or eternal bliss.

Tolstoy did not develop his views with the rigor of a professional philosopher, nor did he propose anything original. Yet he had a practical application many will find appealing. Religious faith allowed him to continue to live rather than wish to die. Believing in God, even against reason, provided the meaning he wanted.

Declining to Believe in God

The emphasis in this chapter has been on those thinkers who believed in God but disagreed on whether or not God's existence could be proved by reason. Some said yes, and some said no. Some other philosophers have thought that God's existence cannot be proved and have decided not to believe in him.

Albert Camus (1913-1960), a Frenchman who won a Nobel Prize in literature, agreed with those who thought the existence of God cannot be proved by reason but did not agree with belief in God. Regarded as an existentialist, he thought he had the freedom to decide for himself. He chose not to believe in God. Camus gave his views in *The Myth of Sisyphus and Other Essays*.

Camus said there were temptations to follow religions or prophets by a kind of leap. He apparently had heard something about Kierkegaard. Yet Camus believed such things are not fully understood. Man seeks to understand but is met by the unreasonable silence of the world. The situation is absurd. The absurd man is the one who is aware of the ridiculous situation and tries to deal with it. Going only by what is understood, the absurd man finds that nothing is certain.

Camus thought that life has no meaning. The problem of life is symbolized in the ancient Greek myth about Sisyphus, who had to roll a large stone up a hill only to have the stone roll down to the bottom—a process repeated over and over without end. For Camus, the story can symbolize life as hard and repetitious with no apparent meaning. The only meaning he found was rebellion against the lack of meaning of life. He advocated courage in a struggle that could not be won, thinking it was noble to continue to fight when there was no possibility of victory.

We can sympathize with Camus on the difficulty of understanding. Who fully understands man and the world? We can also admire his call for courage not to give up even in a hopeless struggle. But the conclusions of Camus, that God does not exist and that life has no meaning, are bleak. The other thinkers covered in this chapter provide much more hope.

Conclusion

Biblical writers thought they had special revelations from God, and then wrote their beliefs in the books that became the Bible. Philosophers usually take a different approach. They have tried to use reason apart from the Bible in thinking about God.

It is intriguing but also puzzling that sincere and brilliant men disagree over trying to prove the existence of God by reason apart from the Bible. Anselm, Thomas Aquinas, and William Paley had different arguments but believed reason could prove God's existence. Pascal, Kierkegaard, and Tolstoy had varying approaches but believed reason could not prove God's existence. The latter group still chose to believe in God, however. Then there are sincere and brilliant men such as Camus who have thought God's existence could not be proved and have declined to believe in God.

How can sincere and brilliant people come to differing conclusions on trying to prove the existence of God by reason? How can they disagree on whether or not to believe in God? Whatever the explanation may be, we are challenged to arrive at our own informed and good decisions.

Questions for Further Thought

1. Why do you think some people emphasize reasoning about God from the Bible and others emphasize reasoning about God apart from the Bible?
2. How would you explain what you consider to be the strongest argument from reason for God's existence?

3. What do you think of Pascal's advice that people should act as if they believe in God even if they have doubts?
4. What might you say to people such as Camus who do not believe in God?

Note

[1]For appropriate selections from philosophers considered in this chapter, along with additional background information, see *Classic Philosophical Questions*, 8th ed., ed. James Gould (Englewood Cliffs, NJ: Prentice-Hall, 1995).

CHAPTER 7

The Nature of God

In addition to trying to prove the existence of God by reason apart from the Bible, another important topic in philosophy has been the nature of God.

Some philosophers have included various accounts of God's nature in attempts to prove God's existence. Anselm set forth the nature of God to some extent by thinking of him as that being than which nothing greater can be conceived. Thomas Aquinas thought of God's nature in relation to experiences of the world. God was First Mover, First Cause, Necessary Being, Maximum of Goodness and Being, and Intelligent Designer. William Paley had his own version of one of Aquinas' arguments with the view of God's nature as the intelligent designer for the world.

For those we have considered who thought that God's existence could not be proved but still believed in God, there were also views of God's nature. Pascal thought that God, if God does exist, would be of such nature as to reward those who believed in him. Kierkegaard was not clear on what God might or might not do, but the Danish writer believed God's nature was beyond rational comprehension. Tolstoy believed that relating to God as infinite in nature would provide significance for finite human life.

Besides views of God's nature by individuals, philosophers have considered basic interpretations of God's nature. These interpretations include traditional theism, pantheism, God as limited, and God as part of the world.[1]

A significant issue for the nature of God in traditional theism is the conflict between the goodness attributed to God and the existence of evil and suffering in the world. Thomas Aquinas referred to the problem when he covered objections to believing in God's existence. David Hume has challenged God's goodness, and John Hick has defended God's goodness.[2]

Traditional Theism

The most common understanding of God's nature is traditional theism, the basic view accepted in Judaism, Christianity, and Islam. According to this monotheistic view, there is one God and that God created the world.

Two scholars who tried to express traditional biblical teachings about God in philosophical terms were Philo of Alexandria and Moses Maimonides. Philo of Alexandria (around 20 B.C.-A.D. 50), also known as Philo Judaeus, tried to relate the God of the Hebrew Scriptures to Greek philosophy. He thought of the creator as being supernatural, perfect, and all-powerful. Moses Maimonides (1135-1204) attempted to connect God in rabbinic Judaism with the philosophy of Aristotle. In his *Guide for the Perplexed*, Maimonides set forth thirteen principles of the Jewish faith. The principles concerning God cover God's existence, unity, incorporeality, eternity, and omniscience.

Christian philosophers have generally come to think of God's nature as having both metaphysical and moral attributes. In philosophy, metaphysics is the study of reality. The metaphysical attributes of God refer to God's basic nature and include omnipotence, omniscience, and omnipresence. These words express belief that God is all-powerful, all-knowing, and all-present or somehow present in all places. The moral attributes refer to God's nature in the sense of what values God has. The moral attributes of God include goodness and love.

In traditional theism, God's relationship to the world is a special consideration. God is thought of as both transcendent and immanent. The transcendence of God refers to God's separateness from the world in being both above and beyond the world. Yet the immanence of God refers to God's also being somehow within and throughout the world. How could God be both transcendent and immanent? Christian philosophers acknowledge that some matters remain mysteries.

Traditional theism is based on the Bible. Some of the words used about God by religious philosophers are in the Bible, but other words are not. Whether or not the words are in the Bible, traditional theists intend the words to express biblical beliefs about God's nature.

Pantheism

Some philosophers have disagreed with traditional theism. One variation from traditional theism is pantheism, or the idea that there is one God but he is all there is; that God is everything or that everything is God. Pantheism is literally "allGodism." Instead of God's being all-this or all-that, God is absolutely everything. God is the world or, the same thing, the world is God.

The most famous pantheist in Western philosophy was Baruch Spinoza (1632-1677). According to Spinoza, God not only is omnipresent or somehow present in all places but also *is* everything everywhere. All things are divine because everything is God. Pantheists do not think they need to try to prove God's existence. They believe that everything people accept as existing is actually God. It is customary to refer to Spinoza as "the God-intoxicated philosopher."

Paul Tillich, a German-American theologian of the twentieth century, had a special way of referring to God. He spoke of God as the Source of Being or the Ground of Being or Being-itself. To call God the Source of Being or the Ground of Being sounds like a philosophical way of saying that God is the creator. To designate God as Being-itself could be understood as suggesting pantheism. In spite of a possible misinterpretation, Tillich was probably intending to say something such as statements expressed by Anselm and Thomas Aquinas. Considering God as Being-itself would not have to be pantheism but something like belief in God as necessary being.

Traditional theism and pantheism differ greatly concerning the nature of God. According to traditional theism, God is the creator of the world and is above the world. According to pantheism, God and the world are the same. According to traditional theism, God is separate from the world so there is a distinction between the sacred and the secular. According to pantheism, there is no distinction between the sacred and the secular because God and the world are the same.

God as Limited

Some scholars have agreed with much of traditional theism except the part about God's omnipotence or being all-powerful. Is God perhaps limited? The idea of a limited deity goes back as far as Plato's dialogues. Edgar S. Brightman, a twentieth-century American philosopher, also had this view.

Brightman believed the problem of evil called for an adjustment in the understanding of God's nature. He thought the existence and persistence of evil required the idea of God as powerful to an extent but not all-powerful. God's goodness impels him to fight against evil. Since much evil continues in the world, there is something wrong. Perhaps God is not able to overcome evil completely. Therefore, Brightman's thinking went, God is limited.

Brightman's view gave God credit for goodness and for doing his best to fight evil. But there is the reluctant conviction that God does not have the power to eliminate evil entirely. This view can be stated in a special way. If God could completely eliminate evil, God would. If God has not completely eliminated evil, God must not have all the power needed. Therefore, the view is, God must not be all-powerful.

Followers of traditional theism have difficulty trying to explain how God can be both good and all-powerful when evil abounds and continues. But traditional theists think the idea of God as limited is not the correct answer.

God as a Part of the World

Traditional theists have thought of God as both transcendent and immanent, as both above the world and within the world. How can God be both transcendent and immanent? Traditional theists are not sure they have a complete explanation, but they believe it is God's nature to be both.

Some have rejected God's transcendence, or God's being above and beyond the world. They think God is only immanent, somehow completely within the world; that some part of the world can be considered divine.

Among modern thinkers, Henry Wieman had difficulty with the idea of God as creator of the world. Wieman did not think of God as separate from the world. He valued creativity, recognizing creativity within the world as being God.

John Dewey, famous in American education, did not think of God as being above the world and directing its progress. Although he considered progress to be very important, he thought God might be understood as progress within the world toward ideal goals. Reaching the goals would be wonderful, but progress in trying to reach the goals would still be commendable. Dewey believed the proper kind of progress by humans would be divine.

Those who believe God is a part of the world are not concerned about trying to prove God's existence. Valuable human activities do occur, and at least some of those activities have been considered to be God or godlike. Traditional theists reject such thinking. They do not accept a reduced view of God, but continue to believe that God is both above the world and within the world.

Hume's Challenge to God's Goodness

While some scholars have questioned the claim of traditional theists that God is all-powerful, Scottish philosopher David Hume (1711-1776) in *Dialogues Concerning Natural Religion* challenged the belief in God's goodness. Hume did not base his concerns on the violence attributed to or allowed by God in the Bible, but questioned the goodness of God in relation to the suffering in the world. He did not elaborate on the suffering people bring on themselves, but dealt with natural evil—or suffering that may be considered due to God as creator.

Hume thought there were four circumstances, when combined, that explained the natural evil in the world. These circumstances might be compatible with a good God but would never prove a good God. Hume claimed we would not immediately think that a good being was behind the circumstances.

According to Hume's first circumstance, all beings are capable of feeling both pain and pleasure. On the assumption that God created beings in this way, we could ask why. A possible explanation is that pain may be helpful toward self-preservation. The pains of hunger, thirst, and fatigue alert us to our needs for food, drink, and rest. Hume wondered if we could not have gotten along just as well without any pain. Could not less pleasure indicate our need for food, drink, or rest? Hume's implied question here is why a good God would make us capable of feeling pain at all.

In Hume's second circumstance, the general laws of the world often make us feel pain. Being capable of feeling pain would not be a problem by itself, but the laws of the world bring pain into existence. Although Hume did not give examples, we can imagine what he may have had in mind. Perhaps it would be something such as falling hard to the ground and experiencing pain. If we had no "law" of gravity, perhaps we would not fall hard and would not feel pain. Hume was bothered by why a good God would establish such laws. Would not a good God have made laws of nature so that we do not experience pain? Why did God not protect us?

Hume's third circumstance for explaining natural evil is the limited ability to avoid painful situations. All beings have some ability to avoid pain. An eagle has wings, a deer has speed, and an ox has great strength. Yet no being has the qualities to avoid misery altogether. Hume thought we might wonder why a supposedly good God would not have provided more protection from pain.

The fourth circumstance in Hume's analysis of natural evil is his claim of inaccurate workmanship in nature. Here Hume appears to challenge the story in Genesis 1 of the goodness of God's creation. Hume thought that nature, supposedly God's creation, displayed insufficient concern for human welfare and brought pain. Humans need, for example, wind and rain and heat. Often we have either too much or too little of these elements of nature rather than exactly the right amount.

We can imagine what Hume did not directly express. Too much wind is evident in ferocious storms, while too little wind would leave sailing ships dead in the water. Too much rain results in devastating floods, but too little rain allows drought conditions. Too much heat is present in scorching deserts, and too little heat can bring suffering in frigid regions.

Hume noted that extremes of wind and rain and heat can bring misery and ruin to individuals. If God is good, why did God not provide more accurate workmanship in nature for the benefit of humans? Hume's implication is doubt concerning the goodness of God expressed in traditional theism.

It can be startling when different persons have widely varying interpretations of what we may consider the same subject. How do people, for example, view the world? Much depends not only on the world itself but also on how we view the world and what we choose to emphasize. Thomas Aquinas looked at the world and thought that certain characteristics of it helped to prove the existence of God. Hume looked at the world and thought that certain characteristics of it helped to explain pain. He had doubts that the pain in the world would make us think of a good God.

Hick's Defense of God's Goodness

John Hick, writing in *Philosophy of Religion* (1970), argued for the goodness of God in spite of the problem of evil. Hick did not believe that he had the entire answer but perhaps some of it. He hoped his views would keep people from not believing in God.

Hick acknowledged that the fact of evil is a problem for many in believing in God. How can the problem, which goes back many centuries, be stated? If God is good, God would wish to overcome evil. If God is all-powerful, God would be able to overcome evil. If God is both good and all-powerful, then God would be willing and able to abolish evil. Yet evil continues to exist. There is a problem.

What should we think? Perhaps God is good but not all-powerful. God might like to overcome evil but may not have sufficient power to accomplish the task. This idea is held by those who, like Brightman, believe God is limited. Perhaps God is all-powerful but not entirely good, at least in relation to human welfare. Could God overcome evil but not care much about helping people? Hume was one who raised doubts about God's goodness concerning human welfare. Hick thought some people might even conclude that God does not exist. We would then have no one to help us in the struggle against evil. We would be on our own.

Hick rejected several possible solutions as unacceptable to traditional Judaic-Christian faith. First, he disagreed with the view of Christian Science that evil is an illusion; he did not doubt the reality of evil. Second,

Hick did not accept the thinking of the Boston "Personalist" School as represented by E. S. Brightman. In this limiting view, God is good but not all-powerful. Hick held to the belief of traditional theism that God is both good and all-powerful.

Third, Hick did not agree with what he called the classic Christian teaching, which was first developed by Augustine (354-430), that evil is real but is not due to God. We should, according to this teaching, regard evil as the distortion of something good. Augustine believed we should not blame God for evil. Hick rejected this view because he did not think it was compatible with belief in the sovereignty of God. How could God be sovereign or all-powerful without having responsibility for the existence of everything, including evil? Did God not start it or at least allow it? Hick suggested that evil is somehow a part of God's plan.

Hick gave his own theodicy, which is an attempt to explain God's goodness in relation to evil. Hick wrote of two kinds of evil: moral and non-moral. He tried to explain God's goodness so as to account for both kinds of evil.

Moral evil is due to human wickedness, what humans bring upon themselves and one another. Hick believed that God did not make people wicked but provided them with freedom. While freedom is a good gift, we can use it for good or evil. God should not be blamed if we misuse our freedom to do bad things. But could not God have made persons who do only what is right? For Hick, that situation was impossible. Being a person involves having the freedom to do wrong.

Why would God give people freedom when he knew they would sometimes use it for evil? The traditional answer is that God gave people freedom so they would be able to choose a relationship with him. A forced relationship would not be a proper relationship. We cannot, however, be free to choose a relationship with God without also being free to choose evil.

The second kind of evil—non-moral evil—is suffering caused by nature, not by humans. Hick considered here what Hume referred to as natural evil. Floods, droughts, storms, and earthquakes can bring great suffering but are due to nature, not to persons. We cannot explain acts of nature by human freedom. We may say that human freedom is a factor in preparing for acts of nature and in responding to them but not in causing them. Who is responsible for nature? Is it not God? Hick concluded that nature, including its evil, must represent something about God's wishes.

How could evil of any kind be a part of God's plan? Hick suggested that God did not intend the world to be a permanent hedonistic paradise.

We should not expect to live in a state of total and undisturbed bliss. God planned instead a place where humans could grow and develop in the deepest and fullest sense. To arrive at such a goal, people must go through some amount of suffering, which might seem evil. A world with no suffering would not allow much emotional and spiritual growth. Hick thought that evil as suffering was consistent with God's goodness in seeking spiritual maturity for Christians.

Hick's idea of growth and deepening through suffering is similar to the view that an artist must suffer in order to produce great art. If the artist has no experience of pain, the work of art might lack depth. There is also the view of athletes and athletic trainers that great muscular development can be achieved only through rigorous exercise. We hear athletes speak of no pain, no gain. Hick had a further interpretation of what Thomas Aquinas said, that evil is real but that God can bring good from it.

Hick had a special word for Christians. He wanted them to react to suffering not with doubt and despair but with positive spiritual growth. Referring to the extraordinary example of the crucifixion of Christ, believers can see Christ's death paradoxically as both the worst and best possible thing. The crucifixion brought the death of God's Son—the worst possible thing. At the same time, Christ's crucifixion was the means God provided for people's salvation—the best possible thing.

Hick believed that Christ's crucifixion allows Christians to see the evil of suffering both negatively and positively, and that Christians should acknowledge the pain of suffering—the negative part—but not dwell on it. Christians should look beyond the suffering to see God's positive and good goal of spiritual growth for persons.

Hick regarded his views as a partial—not the full and final—answer to how God may be considered both good and all-powerful in spite of the evil of human suffering. He had a good explanation for suffering that leads to spiritual growth, but his answer is not as convincing when there is intense suffering over a long period of time. There can be extremely difficult times in the lives of people when it is hard to find spiritual benefits. Such circumstances can be a great challenge to faith.

Hick believed, however, that there must be another life with enough good to make up for all the bad of this life, a life in which God's goodness and power would be fully displayed. While there may be difficult problems in this life, the hope is that God will do away with all of the problems when people come to the next life.

A Challenge for Hick's View

The evil of human suffering continues to be a difficult issue for many people. Perhaps God does allow some suffering for spiritual growth, as Hick indicated. But, if God is both good and all-powerful, how do we deal with extreme situations? Why does God not set a limit on suffering? Why does God not intervene in some heartbreaking cases?

A classic example of the intensity of suffering comes from a story told by the Russian writer Fyodor Dostoevski (1822-1881) in *The Brothers Karamazov*. The story, supposedly based on an actual incident, may make us wonder about the adequacy of Hick's view.

Ivan and his brother Alyosha, a monk, were discussing the existence of God. Ivan recommended there be no thought about the matter. The human mind, he believed, is unable to deal with such issues. Ivan stated that he accepted God simply, but Ivan added that he did not accept the world created by God. Ivan was concerned about suffering, especially the suffering of children. In explaining his views, Ivan related a shocking story about the suffering of one child.

There was once a retired general, an owner of great estates who had absolute power over the lives of all those under him. One day an eight-year-old serf boy threw a stone while playing. The stone accidentally hurt the paw of the general's favorite hound. Upon learning what had happened, the general assembled the child, the child's mother, and others outside on a cold, freezing morning. The general had the child stripped naked and then set the dogs on the child to tear the boy to pieces before everyone. The general was later declared incompetent.

Ivan did not think the suffering of even one innocent child could be properly explained. He thought that shooting the general would hardly make up for all the suffering. As Ivan considered the matter, he did not reject God but thought he would like to return his ticket to the world. The suffering of children in the world was more than Ivan thought he could endure.

Would philosophy, even religious philosophy, have helped Ivan? Perhaps he could have considered the idea that God provided the good gift of freedom but that people can misuse freedom in horrible ways. Is the suffering of children too great a price to pay for having freedom? Could not God allow freedom to choose (or reject) him without allowing some people freedom to do horrible things to children? Would Ivan have accepted the idea of spiritual growth through suffering? It is hard to see how the child in Ivan's story could have experienced spiritual growth. The child was killed. Would the mother of the child have experienced spiritual growth? She may have eventually, but we could understand if she first experienced devastation. Would others at the scene have experienced spiritual growth after witnessing the general's cruelty? These questions are all very difficult.

The proper response to suffering from those who hold to traditional theism is a great challenge. How could God's nature be both good and all-powerful when there is so much suffering in the world? Traditional theists such as Hick may not now have a complete, final answer to all of the questions that may be raised. But would it be better to have no explanation? Those who hold to traditional theism can claim at least some explanation as to how God can be both good and all-powerful in spite of human suffering. Traditional theists also hope that God will provide for greater understanding and a better existence in another life.

Conclusion

Philosophical thinkers have given much attention to the nature of God. Some included beliefs about God's nature when they were considering whether or not God's existence could be proved.

General views of God's nature have included traditional theism with its claim that God is both all-powerful and good. Another view of God's nature is pantheism, the belief that everything is God or God is everything. Because of the existence and persistence of evil, some have the view that God's nature is limited, that God may be good but not all-powerful. Then there is the view that God's nature does not include transcendence, or being above and beyond the world. In other words, God is an important and valuable part of the world, like creativity or progress toward desirable goals.

Traditional theists are the largest of the groups with these views, but they also have difficult problems, one of which is trying to reconcile belief in God's goodness and unlimited power with the evil in the world. David Hume saw circumstances of pain in the world and challenged belief in God's goodness. John Hick defended God's goodness by emphasizing God's plan of spiritual growth through suffering, and claimed he had only a partial explanation and looked for further answers in the next life.

Questions for Further Thought

1. How would you explain which of the various views of God's basic nature you consider to be correct?
2. How would you express your agreement about God with Hume or Hick or both or neither?

3. What is your explanation for evil in the world, especially in view of the belief within traditional theism that God is both good and all-powerful?
4. What might you say about God to someone such as Ivan, who was distressed over the suffering of a child?

Notes

[1] Although many interpretations of God's nature are generally known by philosophers and religion scholars, the presentation in the first sections of this chapter for others is largely an adaptation of some of the material in chapter 7 of *Invitation to Philosophy* by Stanley Honer, Thomas Hunt, and Dennis Okholm, 7th ed. (Belmont, CA: Wadsworth Publishing Co., 1996).

[2] For appropriate selections from Hume and Hick, see *Classic Philosophical Questions*, ed. James Gould, 8th ed. (Englewood Cliffs, NJ: Prentice-Hall, 1995).

PART 3

***

God and Science

CHAPTER 8

God and Time

Having given some attention to God and the Bible along with God and philosophy, we come now to God and science. As we think about God and science, we do not omit consideration of the Bible as philosophers often do. Our thinking will involve beliefs about God based on the Bible in relation to the scientific method, scientific knowledge, and scientific theories.

The basis for science is observation by means of the five physical senses, which are often enhanced by various instruments. The scientific method involves reasoning on the basis of what has been observed.

How can scientists think about God since God as Spirit cannot be observed? Scientists in earlier centuries were usually willing to consider the possibility that God was behind much, if not all, of what scientists observed. There was the thought that God as invisible was responsible for visible effects. God could not be directly observed, but God's works could be observed.

Modern scientists are much less likely to accept this earlier way of thinking about God. They look for natural causes of natural effects and generally do not go beyond those considerations. Individual scientists may believe in God, but most scientists today do not think that God falls within the area of science. The prevailing view is that it is inappropriate to discuss God in a science course.

How then can there be thinking about God and science? Under current circumstances, the best available procedure may be to consider biblical beliefs about God in relation to scientific discoveries and theories. How do biblical beliefs about God compare to scientific views of the world? While all of science is large and complicated, we will limit our considerations to time, life, and space, which themselves are enormous topics. The goal is not to consider everything about these areas but to examine some of the major views in religion and science concerning them.

We turn in this chapter to a consideration of God and time. We will consider the nature of time, a religious interpretation of the extent of time, and a scientific interpretation of the extent of time. Then we will think about how close or far apart religion and science are regarding God and time.

The Nature of Time

If we wish to think carefully about time, we might consider a book such as internationally famous Stephen Hawking's *A Brief History of Time*. Will this book help us to have an adequate scientific understanding of time? Contrary to what we might expect, the book does not deal much with the nature of time in the ordinary sense. Nor is there a chronological sequence from beginning to end that is usually associated with a history. While Hawking does have references to time, he also covers topics such as temperature and gravity in the universe, Einstein's theories of relativity (both general and special), quarks, black holes, and the accomplishments of various scientists who have won a Nobel Prize in physics. It appears that Hawking had much more in mind than time and does not devote as much space to it as we might have anticipated.

We can get a better idea of his intentions by looking at some of his other works. In his brief foreword to *The Universe in a Nutshell*, Hawking displayed a playful, almost sophomoric sense of humor, which at least some of us can appreciate. Replying to people who asked him about a possible sequel to *A Brief History of Time*, he said that he did not wish to write *Son of Brief History* or *A Slightly Longer History of Time*. He was busy with research and thought that another kind of book would be easier to understand.

What then was Hawking thinking when he wrote *A Brief History of Time*? It is helpful to look at *Stephen Hawking's A Brief History of Time: A Reader's Companion*. In the first sentence of the foreword, Hawking states his main aim is to tell about progress being made in understanding the laws of the universe. We might think another title, perhaps one referring to the laws of the universe rather than only to time, could have expressed that aim in a clearer fashion.

Probably most of us think of time from the viewpoint of common sense. Whatever astrophysicists may have to tell us about special understandings of time, most of us think of time in a fairly ordinary way. Time is one of those concepts we often take for granted. Everyone supposedly knows what time is even if it is difficult to put into words.

The concept of time being considered here is the measure of change. Discussion of this idea of the nature of time goes back at least as far as Augustine of Hippo. Things change—some quickly and some slowly. Noticing how things change and how quickly or slowly gives us our sense of time. If nothing ever changed, there would be no time. If everything were in a state of suspended animation with no movement, there would be no change and thus no time. If we wish to express the view that absolutely nothing whatever happened, we might use the figurative expression that time stood still. In order for change to occur, there must be something capable of change. Material things are capable of change that can be observed, so we calculate time by observed changes in material things.

What changes in material things do we use for calculating units of time? A natural unit of time is a day. The sun appears to rise, move across the sky, and then set. The process is repeated over and over again. We often think of a day as sunup to sundown, the hours of daylight. We also think of a day as the period for the sun to return to its previous location in the sky, such as one sunrise to another sunrise, when a new day begins. Jews think of a day as one sunset to another sunset. Astronomers consider a day to be one complete revolution of our planet on its axis.

A larger natural unit of time is a lunar month. In language used more frequently in earlier centuries, the moon appears to wax (get larger) and wane (get smaller). The moon does not change size, but we see different amounts of light reflected from its surface. We can observe the moon's going from a new moon (smallest size) to a full moon (largest size). From our perspective, the period from one new moon to the next new moon is slightly more than $29^{1}/_{2}$ days. It made sense for largely outdoor people, such as Native Americans, to tell time by the number of full moons from a particular event. Something could be described as having happened so many (full) moons ago.

Lunar calculations continue to be important in various religions. Christians in the western world always celebrate Christmas on December 25, but the celebration of Easter varies from year to year according to the moon. The date for Easter is the first Sunday after the first full moon after March 21, the spring equinox (when daylight and dark are equal). Muslims observe their fasting month of Ramadan at different times from year to year because of calculations based on a full moon.

A third natural unit of time is a year. In most of the world there are four distinct seasons. There are dramatic changes with spring, summer, fall, and winter. Since the sequence of seasons occurs over and over, we can understand a year as the length of the four seasons or as the period from the beginning of a season until it begins again.

Another way of thinking about a year is to consider the amount of daylight in relation to the amount of darkness in days. We think of longer days with more duration of sunlight compared to the shorter days with less duration of sunlight. The period from one longest day with the greatest duration of sunlight to another longest day, or one shortest day to another shortest day, is a year. Some ancient peoples, including those who built Stonehenge, probably made such calculations.

There is also a solar year. When astronomers speak of a solar year, they think of one complete orbit of our planet around the sun. The length of the orbital period is approximately $365\frac{1}{4}$ days. Our calendars have three years of 365 days each and then a fourth (leap) year of 366 days. Since a solar year does not have an exactly even number of full days, our calendars are always very slightly inaccurate even with leap years.

Although there are various natural units of time or ways to measure change, there are additional types of measurement. Dividing a day into 24 hours, an hour into 60 minutes, and a minute into 60 seconds is not natural but is convenient. People have calculated time with such manufactured objects as sundials, sand (hour) glasses, water clocks, pendulum clocks, wristwatches, and atomic clocks. Mechanical ways for calculating time took on some personal features as people referred to the "face" and "hands" of a clock or watch. We have much less of a personal reference with digital timepieces.

What is the point of covering information about time that most of us think we know? The idea is that time is more complicated than we sometimes think it is. Even when we do think that time is complicated, it may be even more complex.

Time becomes extremely complicated when we consider the far reaches of space or the distant past. Since our views of time, especially days and years, are based primarily on the sun, what should we think about time beyond our solar system? Is it still appropriate to speak of days and years when we consider other solar systems and even other galaxies? Perhaps we cannot go much beyond what is familiar to us, but the universe is very big. There are many places in the universe where our sun, the focal point of time for us, seems insignificant.

Also, what should we think of time before our solar system existed? Before our planet and our sun and our moon came into existence, what was the meaning of time and how could time be calculated? What change allowing for time, if any, was there then? Did time itself have a beginning? If so, what was there before time?

The situation does not get easier when we think about God and time. What does time mean to God? According to one viewpoint, God is greater than time and therefore infinite and beyond time. According to another understanding, God is involved with time and somehow influences events on our planet and elsewhere in the cosmos according to his schedule.

Some people believe both of these viewpoints may be true. We may think that God is beyond time but also works within time. How both views can be true is a mystery.

From the viewpoint of common sense, we think we have a good understanding of the nature of time. Thinking of time as the measure of change sounds like a reasonable idea. But our ways of measuring change may be very limited.

A Religious Interpretation of Time

When we think about time, a big question among many others is when the world came into being. How long ago was that? Various religious interpreters have tried to answer the question by studying the Bible.

A famous religious interpretation of time came from Archbishop James Ussher (1581-1656). According to his calculations, creation occurred in 4004 B.C. Ussher's dates for events in the Bible, including creation, are in the margins of some Bibles, especially older Bibles.[1] It was Dr. John Lightfoot of the University of Cambridge who specified in the seventeenth century that creation occurred on October 23, 4004 B.C., at nine o'clock in the morning.[2]

In spite of such additions as those made by C. I. Scofield in his edition of the Bible, we do not find in the Bible text itself dates with such calendar designations as B.C. and A.D. (or B.C.E. and C.E.). We do, however, find information about time in the Bible, information that may be used in calculating the date for creation of the world.

The story of creation in Genesis 1 refers to six days of creation by God with a seventh day of rest. But, according to the Bible, how long ago did those events occur? There are various places in the Bible that indicate periods of time.

Genesis 5 lists the descendants of Adam, the first man. These people lived before the time of the great flood in the days of Noah and are reported as having exceptionally long lives. According to verse 1, Adam lived 930 years. Methuselah lived even longer: 969 years (v. 27). There is much information in this genealogy about how long various men lived and how old they were when various sons were born. This information can be used to calculate time from the creation of Adam through many generations.

Another way for calculating periods of time in the Bible is to go by the length of the rule of kings. In the books of 1 and 2 Samuel, 1 and 2 Kings, and 1 and 2 Chronicles, there are indications of the ages and length of reigns of the kings of Israel and Judah. For example, we read in 2 Chronicles

28.1, "Ahaz was twenty years old when he began to reign; he reigned sixteen years in Jerusalem." This kind of information is often given in the same form for other kings. According to the information about the length of time that the various kings ruled, there were Jewish kings for more than four centuries.

If we would like to have a very extensive indication of time in the Bible, we may turn to the genealogy in Matthew 1. According to verse 17, "So all the generations from Abraham to David are fourteen generations; and from David to the deportation to Babylon, fourteen generations; and from the deportation to Babylon to the Messiah, fourteen generations." There may be various interpretations of the exact number of years in a generation, but the genealogy provides a basis for thinking of a large amount of time from Abraham to the Messiah.

For an even longer genealogy in the New Testament, we may look at Luke's Gospel. He provides a list of names from Jesus all the way back to Adam. (See 3.23-38.) There is no indication of the total number of years or of the time between each father and son, but it is still possible to estimate a large amount of time from Jesus back to Adam.

Whatever information Ussher used, a literal interpretation of the Bible provides material for calculating time from creation. Ussher's date of 4004 B.C. for creation has been very influential as a religious interpretation of time.

A Scientific Interpretation of Time

When scientists think about time, they go beyond but still include ordinary thinking about time. They accept time as the measure of change in material things as one perspective. Such additional scientific views as a space-time continuum, however important such considerations may be to astrophysicists, are not our concern at present. Let us consider how a scientific interpretation of time in the ordinary sense differs from a religious interpretation.

The scientific interpretation of time differs from Ussher's religious interpretation of time concerning the amount or extent of time from the beginning of the universe. In opposition to the view of Ussher that the world was created a little more than 4,000 years before the birth of Jesus, most scientists today contend that the world is much, much older. There is also the belief among scientists that the universe started much longer ago than our planet did.

We may read claims by archaeologists of evidence for the beginnings of great ancient civilizations as long ago as 8000 B.C., about 4,000 years before Ussher thought the world was created. We may consider the views of anthropologists that there are indications of the existence of modern humans at least 30,000 years ago. The truly startling numbers come from the astronomers or, as they are increasingly being called, the astrophysicists.

If we go back to Stephen Hawking's *A Brief History of Time*, we find his view of the beginning of the universe. Without being too precise, Hawking thought the universe was at its minimum size ten or twenty thousand million years ago.[3] (It is not clear why he used thousand million rather than billion.) Hawking reflected the view now widely accepted among scientists that the universe began with a Big Bang of extremely dense material. After a period of expanding, cooling, and other activities, the universe formed various bodies. According to this view, our planet came into existence about five billion years ago.

If we would like somewhat more precise figures, we may turn to *Origins: Fourteen Billion Years of Cosmic Evolution* by Neil deGrasse Tyson and Donald Goldsmith. These scientists explained their views of the origin of the universe, the origin of galaxies, the origin of stars, the origin of planets, and the origin of life. They claim that the universe (and time) began 14 billion years ago.[4] They later more carefully state that the beginning of the universe was 13.7 billion years ago.[5] The authors say that astrophysicists have no idea what came before the beginning of the universe.[6]

How do scientists calculate age? If they do not have witnesses or documents, as historians do, scientists rely on their standard method of observation. Consider the simple example of the age of a tree that has been cut down. Scientists, as well as others, can observe the number of circles or rings in the trunk of the tree. Each ring represents a year's growth. It is usually easy to count the number of circles or rings in the trunk and determine the age of the tree.

What is the age of the universe? How can scientists come up with a number? There are various considerations, but one way to think of the age of the universe is to think of the size of the universe. How long did it take for the universe to become as big as it is? A widely held earlier view was that of the "solid" state of the universe, that the universe was not getting any bigger and not getting any smaller. Then Edwin Hubble made astronomical observations in the late 1920s that led him to conclude that the universe was expanding. The view of the universe as expanding is the prevailing scientific view today.

What happened before expansion of the universe began? The current scientific belief is that the material of the universe before expanding was once extremely small and dense. The material then somehow exploded in what is called the Big Bang. How long ago did that happen? Calculating the rate of expansion of the universe is not as simple as counting the rings or circles in a tree trunk. But the current scientific estimate already mentioned is that the universe is 13.7 billion years old. That figure represents the approximate amount of time that scientists believe it took for the universe to become as big as it is today.

The dominant scientific interpretation of the age of the universe differs considerably from the religious interpretation of Bishop Ussher. There is a huge difference between 4004 B.C. and 13.7 billion years ago for the beginning of the universe.

Different Interpretations of Time

Must we choose between the scientific interpretation of time and the religious interpretation of such people as Ussher? Is there the possibility of change in either interpretation?

There may be some adjustments in the scientific interpretation of time due to new knowledge. But it is not likely there will be much change in the scientific interpretation of the amount of time since the universe began. From the viewpoint of observation, the scientific interpretation of time is well established. A religious interpretation of time like that of Bishop Ussher will have virtually no scientific credibility for almost all scientists. A view of time based on ancient writings might have historical value for scientists in indicating how people long ago used to think. Even though many people today continue that way of thinking, most scientists believe they have a far better method than the Bible for gaining knowledge of the world.

Many who hold a religious interpretation of time are unwilling to change their views because they believe the Bible contains truth, including scientific truth. Such people may be startled or disappointed or confused or angered by scientific views. They may not be especially concerned about time as such but about what may seem to be an attack on a divine source of truth.

Although religious devotion can be intense, there are questions concerning the proper religious interpretation of time. Have Bishop Ussher and others who agree with him interpreted the Bible correctly regarding time? Let us consider other possible religious interpretations of time.

One possibility that Ussher's interpretation of God and time may not be accurate is how God may think of time. According to 2 Peter 3.8, "But do not ignore this one fact, beloved, that with the Lord one day is like a thousand years, and a thousand years are like one day." (See also Ps. 90.4.) If we wish to be very strict in interpreting this verse, we might say that the six days of creation in Genesis 1 were actually a thousand years each or a total of six thousand years. Such a view would probably not have much appeal for the scientific community but would challenge Bishop Ussher's calculations. The deepest meaning in 2 Peter 3.8 is probably that our way of counting time may not be the same as God's way of counting time—whatever that may be.

A second possibility that there may have been something wrong with Bishop Ussher's thinking is the meaning of "day" in Genesis 1. Ussher and many others have interpreted the days of creation in Genesis as being 24-hour days. It is understandable that we would interpret the Bible in terms of what we know or think we know. But let us look more carefully at the first chapter of the Bible.

The story of creation in Genesis 1 includes the creation of the sun on the fourth day of creation. (See vv. 14-19.) The writer in verse 16 says of the fourth day of creation: "And God made the two great lights, the greater light to rule the day, and the lesser light to rule the night; he made the stars also." The greater light to rule the day has to be the sun, while the lesser light to rule the night is the moon. We determine a day mainly by our planet's relationship to the sun. How could there have been a 24-hour day before the creation of the sun?

Might the first three days of creation, before the creation of the sun, have been something very different from 24-hour days? If those days did not last 24 hours, how long were they? Might the first three days of creation represent extremely long periods of time? If so, there is a possibility of some accommodation between a religious interpretation of time and a scientific interpretation of time. There can be a religious interpretation of time based on Genesis 1 that far exceeds the amount of time proposed by Bishop Ussher.

If we look carefully at the first verses of Genesis, we may become even more concerned about finding the proper interpretation of God and time. We read in Genesis 1.3 of the creation of light on the first "day" of creation. This light is not the light of the sun, since the greater light to rule the day was not created until the fourth day of creation. According to Genesis 1.4, God saw that the light was good and separated it from the darkness. Then in verse 5 we read, "God called the light Day, and the darkness he called Night. And there was evening and there was morning, the first day."

The created light on the first day that was not the light of the sun was called "day." How long was that day? The length of the day is not clear. It is also not clear how there could be evening and morning without a sun. Evening and morning must have been determined by some relationship between light and darkness apart from the sun.

A careful look at Genesis 1 helps us to see that there are weaknesses in Bishop Ussher's interpretation. It is also apparent that there can be more than one interpretation of the creation account. What interpretation is the correct one? The creation account is so profound that it is hard to tell the exact and full meaning.

What other religious interpretations of time, including time from the beginning, might we consider? In his book *Cosmos,* Carl Sagan referred to a three-volume history of the world that is now lost but that once was in the famous library of Alexandria. The work came from a Babylonian priest named Berossus. (Was this person a Jewish priest in Babylon?) In the first volume, the priest set the period between the Creation and the Flood at 430,000 years.[7] Unfortunately, Sagan does not tell where the priest obtained this information or how the priest made his calculations.

A further religious interpretation of time involves the view that the Bible is important but should not always be taken literally in its details. With this view, there could be the conviction that God created the universe but did so according to scientific estimates of the age of the universe. There would be agreement with the biblical belief in God as creator, but there would be openness to scientific views of how old the universe is. Such a view is not acceptable to biblical literalists but does provide a possible combination of religious and scientific concerns.

There are clear differences between religious and scientific interpretations of the amount of time from the beginning of the universe. The biggest difference may be over whether or not God is responsible for the beginning of the universe and thus the beginning of time. Religious interpreters, whatever amount of time they may estimate from the beginning, will say that God is responsible. Scientific interpreters of time probably will say that the question of who or what is responsible for the universe and time is beyond science.

Conclusion

Although there are various interpretations of the nature of time, including a space-time continuum, the ordinary understanding of time is the measure of change that can be observed in material things. If there were no change, there would be nothing to measure and thus no time. Natural units of time include days, phases of the moon, seasons, and years. In addition to natural units of time, there are arbitrary but convenient units of time such as hours, minutes, and seconds that are displayed with various mechanical and electronic devices.

When we think about a religious interpretation of time, we often think of a system associated with Bishop James Ussher. His interpretation, based on his study of the Bible, was that creation occurred only a little more than what would now be 6,000 years ago, specifically in 4004 B.C.

When we think about a scientific interpretation of time, the consensus of most modern scientists is that the universe began about 13.7 billion years ago. Many religious people have objected to the scientific interpretation because of their view that it goes against the Bible or at least the interpretation expressed by Bishop Ussher.

When we give further consideration to the Bible, we find there can be questions about Ussher's interpretation of time, including the view that creation occurred in six 24-hour days. The meaning of "day" in the first three days of creation, when there was created light that was not the light of the sun, is not entirely clear. It may be that the first three days of creation were not intended to represent 24-hour days but much longer periods of time.

An additional religious view is that the Bible, including the story of creation in Genesis 1, is important concerning spiritual matters but should not always be understood literally with regard to science. This view allows for belief in God as creator of the world and time and also permits the greatest amount of openness to scientific concepts of time from the beginning.

While religious and scientific interpretations of the amount of time from the beginning of the universe have usually differed, the most significant difference has to do with God. Is God responsible for time or at least the conditions for time? Religious interpreters would all answer yes. Scientists would probably reply that the question is beyond their area.

Questions for Further Thought

1. What interpretation of the nature of time seems best to you? Why?
2. Do you think Ussher interpreted the Bible correctly concerning time? Why or why not?
3. Which source do you think would be the best way to determine the amount of time from the beginning of the world: the Bible or science? Why?
4. What are some ways, if any, that religious interpretations of time and the scientific interpretation of time might come closer together?

Notes

[1] See, for example, the King James Version as edited by C. I. Scofield.

[2] See Andrew White, *A History of the Warfare of Science with Theology in Christendom*, vol. 1 (Amherst, NY: Prometheus Books, 1993), 9, 256.

[3] Stephen Hawking, *A Brief History of Time*, 10th ed. (New York: Bantam Books, 1998), 143.

[4] Donald Goldsmith and Neil deGrasse Tyson, *Origins: Fourteen Billion Years of Cosmic Evolution* (New York and London: W. W. Norton, 2005), 25.

[5] Ibid., 54.

[6] Ibid., 44.

[7] Carl Sagan, *Cosmos* (New York: Ballantine Books, 1985), 12.

CHAPTER 9

God and Life

As we think further about God and science, an additional subject is life. Among the most majestic but also most mysterious of realities, life presents a challenge to one's understanding. There are ways to describe the nature of life, but the descriptions often lack the vitality of life itself.

There are both religious and scientific interpretations of life, especially regarding the origin and development of life. The two kinds of interpretations often disagree. The basic disagreement is over the place of God in relationship to life. Religious interpretations emphasize God as responsible for life, but modern scientific interpretations do not. How to solve this disagreement or even lessen it is a major problem.

The Nature of Life

Life is difficult to define. From the viewpoint of general observation and common sense, we may say life is the mysterious quality that allows an organism to take in food, grow, and have the ability to reproduce. This account may be accurate but hardly seems adequate. From the viewpoint of chemistry, Tyson and Goldsmith mention that living things on our planet consist almost entirely (99%) of hydrogen, oxygen, carbon, and nitrogen.[1] While not especially inspiring for most of us, these chemical elements are significant for the scientific understanding of life.

These considerations are important, but they do not fully convey the great value life has for living things. Almost all individual living things strongly seek to protect their lives in order to enjoy the benefits life provides for them. There is often great need for protection because of many threats. Enemies of life include disease, accidents, old age, and attacks from other beings. Disastrous consequences for life can come from floods, storms, earthquakes, volcanic eruptions, and meteors that hit our planet. Life is valuable but often fragile.

The two basic biological classifications of life are plant and animal. At the cellular level the two kinds of life are very similar. At a higher level an important difference involves movement from one place to another. Plant life can move by means of growth but cannot escape a particular location on its own. Animal life can not only grow but also is capable of changing locations. In addition, plants and animals consume different types of food.

Plants do not eat other plants. Except for a plant such as Venus' flytrap, plants do not eat members of the animal kingdom. Plants live and grow basically from nutrients in the soil, from water, and from the action of the sun. In contrast, the food of animals is other life. Some animals eat only plants (herbivores), some animals eat only other animals (carnivores), and some animals (like humans) eat both plants and animals (omnivores).

Since animals feed on other life but plants usually do not, we might think that plant life came before animal life. But when did life of any kind begin? When and how did plant life come into existence? When and how did animal life come into existence? When and how did human life come into existence?

We usually find strong disagreement between religious interpretations of life and most modern scientific interpretations of life. Religious interpretations include God as the creator of all life. Most modern scientific interpretations of life, including human life, have little or no reference to God. The main scientific approach is to try to account for life with nothing more than natural explanations.

Religious Interpretations of Life

Traditional religious interpretations of life and God's relationship to it are based primarily on the Bible. The first two chapters of Genesis provide impressive stories.

According to Genesis 1, God created life on the third day of creation (vv. 11-13). God commanded the earth to bring forth vegetation, both plants and fruit trees. On the fifth day of creation (vv. 20-23) God commanded the waters to bring forth living creatures and the birds to fly above the earth. On the sixth day of creation (vv. 24-31) God arranged for other beings, commanding the earth to bring forth all kinds of living creatures, including cattle and creeping things and wild animals. God then made humans, both male and female, in God's image. The humans were to be fruitful and multiply and have dominion over the fish, birds, and animals. God gave green plants and trees with their fruit for food for humans and for animals and birds and creeping things. At this point, animals ate plants but not other animals.

Genesis 1 indicates God created all of life, proceeding from plants to animals of the sea, air, and land. Then God created humans, thus completing the separate groupings of living things.

Genesis 2 gives an additional version of the creation of life. According to verses 4-9, before there was any plant of the field, God formed man (*adam* in Hebrew) from the dust of the ground (*adamah* in Hebrew). It was not until God breathed life into his nostrils that man became a living being. Then God planted a garden in Eden and put man in it. God made trees good for food to grow out of the ground. Verses 18-25 add more to the story. Thinking it was not good for man to be alone, God made animals and birds. Those creatures were not sufficient, however. God then caused a deep sleep for the man, took one of the man's ribs, and made a woman from the rib. The man then recognized the close relationship between himself and the woman.

The second version of creation covers God's bringing into being a human male, plant life (in the Garden of Eden), animals, birds, and then a human female. The account does not specifically mention sea life.

Both Genesis 1 and Genesis 2 emphasize the importance of humans in God's creation. Most important is God, because God as creator is responsible for humans and all other life. There is no indication that any form of life began apart from God.

Scientific Interpretations of Life

Modern scientific interpretations of life concentrate on evolution with little or no reference to God. The basic idea of biological evolution is that different kinds of life forms have come about through natural changes over long periods of time.

When we consider Charles Darwin's view of evolution, widespread opinion concludes he basically believed that humans are descended from monkeys. This idea received wide publicity when John Scopes was being tried in Dayton, Tennessee, in 1925 for teaching Darwin's theory in a public school—a violation of a state law. Scopes was found guilty, but that verdict was later overturned on a technicality.

Many of us have probably seen drawings that try to portray additional ideas of Darwin. The composite picture in my mind begins with a fish that has emerged from water onto land and that has something like a preliminary pair of legs. The fish is followed in succession by a four-legged land animal, a crouching monkey, an upright caveman, and then a modern human. This picture expresses the view that life began in the sea and progressed from fish through animals to modern humans.

What did Darwin say about evolution? In *The Origin of Species* he displayed a vast knowledge of both plant and animal life and geology. Although the book is quite long (almost 700 pages), Darwin frequently indicated he did not have space to tell all that he could. His accumulation of facts is almost overwhelming, his interpretations controversial and not always easy to follow.

Although Darwin was concerned about species, he acknowledged the difficulty of defining the term.[2] He was probably thinking especially about disagreements between naturalists. Since Darwin did not give a clear meaning for the word, it might be helpful to attempt a working definition. For most of us, a general definition of "species" is a group of living things (plants or animals) similar to one another and also different from other groups of living things. But how similar to one another and how different from others do living things have to be in order to be called a species? Naturalists may disagree over whether some living thing belongs to one species or another or is perhaps a "variety" of a particular species.

A common understanding is that individuals of a species can have offspring only with members of the same species. Thus, cats can produce other cats and dogs can produce other dogs, but a cat and a dog cannot have offspring. Although this view of species (reproductive isolation) makes sense to many who are not scientists, not all naturalists are convinced it applies to all groups of living things.

In spite of difficulties with defining the term "species," Darwin challenged the concept of the immutability of species, which is the conviction that species have not significantly changed since creation and cannot change into another species. According to the story of creation found in Genesis 1, God created such life forms as birds, fish, land animals, and humans. In Darwin's day, people generally and most scientists did not see much change in life forms of their time from those mentioned in Genesis. Thus there was the idea of the immutability of species. Darwin (somewhat controversially) stated belief in the Creator, but did not think the findings of science supported the view of the immutability of species. He believed that new and different species had developed from those originally created.

How did Darwin explain his view that new species have come into existence?

He believed that living things have a tendency not only to reproduce but also to reproduce abundantly. Because of such conditions of life as limitation of resources and natural enemies, there is a struggle for existence. Therefore, many individuals do not survive. The tendency is for individuals with superior characteristics to continue to live in what is called "the survival of the fittest" or "natural selection." Human breeders of crops and animals select certain individuals and reject others in what may be termed artificial selection. Darwin contended there is also a natural process by which superior individuals usually continue to live and others do not. Surviving individuals then ordinarily pass on their superior characteristics to their offspring.

Darwin also spoke of descent with modification. How could there be modification? We should think not merely of inheriting superior characteristics from a single parent. There would be a combination of contributions from two parents. That combination would ideally lead to modifications that not only would be new but also helpful in the offspring. Over a long period of time, a different form of life or new species would emerge.

Descent with modification is supposed to occur over long periods of time. Darwin allowed intervals of a thousand or more generations in a series of lines of descent.[3] He also stated that each line could represent a million or more generations.[4] According to Darwin, all organic beings at the dawn of life had the simplest structure.[5] Then higher, more complicated forms of life (new species) developed over an immense number of years.

What evidence did Darwin have for his theory? He referred, among other items of information, to similar bone structure as indicating descent with modification from common ancestors. He found likeness of bone structure in the hand of a man, the leg of a horse, the wing of a bat, and even the fin of a porpoise.[6] He also thought it probable from the similarity of their embryos in their earliest stages that mammals, birds, fish, and reptiles were the modified descendants of an ancient progenitor, which he described.[7] Darwin believed all animals and plants are descended, respectively, from only four or five progenitors.[8] He also thought the fossil record, though far from complete, gave some support to his views.[9]

Darwin had some special comments. In spite of his own vast learning, he admitted there was much ignorance on the origin of species and thus much not yet explained.[10] He also stated that he could not see any good reason why his views should shock anyone's religious feelings.[11]

In *The Descent of Man* (also close to 700 pages), Darwin repeated some of his main ideas but was more specific regarding humans. His evidence for believing that humans descended from a lower form included similarity of bodily structure between men and animals.[12] He found a further sign of such descent in the similarity of embryos of men and animals, especially in the earliest stages.[13] There is also a similarity of rudimentary organs (such as an appendix) in men and animals.[14] These organs may have been well developed at an earlier time but became diminished in later generations.[15]

Darwin tried to clarify some of his views concerning the genealogy of man.[16] He wrote that we should not believe the earlier progenitor of man looked like or even closely resembled any existing ape or monkey.[17] Darwin thought that man may have diverged from the Catarhine stock of monkeys as early as the Eocene period.[18] (The Eocene period, according to geological calculations, lasted from approximately 55,000,000 to 35,000,000 years ago.) Noting that the fossil record is incomplete, Darwin did not see any great difficulty for his view because of the lack of fossil remains connecting humans with ape-like progenitors.[19]

Darwin wrote that evolutionists believe the five great vertebrate classes (mammals, birds, reptiles, amphibians, and fishes) are descended from one prototype. Since the class of fishes appeared first, he concluded that all vertebrates descended from a fish-like animal.[20] Darwin traced the ancient progenitors of vertebrates to marine animals similar to the larvae of Ascidians. Next came fish and then amphibians. Darwin said that no one at the time of his writing was able to say how mammals, birds, and reptiles were derived from fish and amphibians. He found it easier to trace mammals, including placental mammals, through various stages to New World and Old World monkeys. At some remote time, Darwin asserted, man proceeded from Old World monkeys.[21]

Although Darwin expressed his views with considerable learning and sophistication, the general understanding of his basic ideas is close to accurate. He did not say that humans descended from beings similar to monkeys of his day, but he did believe that humans long ago broke off from Old World monkeys. He thought life began with tiny marine animals in the sea and then eventually progressed to fishes and then to mammals and later to humans.

Darwin wrote that his views (including human descent from some kind of monkeys) would be highly distasteful to many. He did not think, however, that there was any doubt that we have descended from barbarians—a group that included savages who tortured enemies, offered bloody sacrifices, practiced infanticide, and treated wives like slaves. Darwin indicated that he would rather think of being descended from some appealing animals he knew about instead of considering being descended from such savages.[22] He thought further that man should have some pride in the great heights to which he had risen, referring to man as the wonder and glory of the universe.[23]

While Darwin was not concerned with explaining his exact religious beliefs, he did write in a positive way about God. He wrote that many wonderful forms of life have come from the few forms or one form originally breathed by the Creator.[24] Darwin also said that the question of the existence of a creator has been answered affirmatively by some of the highest intellects the world has known.[25]

Richard Dawkins, an atheist, informs us that the reference to the Creator did not appear in the first edition of *The Origin of Species* but was used in the subsequent five editions.[26] Perhaps Darwin felt pressure to make some religious comments, but he did publish them in the two books that have been covered here.

Although Darwin dealt primarily with a biological explanation of life, others used Darwin's views to promote their own controversial ideas. Herbert Spencer advocated social Darwinism (let strong people survive and let weak people perish). Francis Galton, a cousin of Darwin, came up with eugenics (let only the best people have children). Thomas Huxley, a strong defender of Darwin's views, argued for atheism.[27] These additional ideas are often associated with Darwin's theory but are not due to Darwin himself.

Another scientific interpretation of life is what might be called the stars-DNA theory. This interpretation of life adds considerations from astronomy and chemistry to the biology and geology of Darwin's views.

Claiming that evolution is a fact and not a "theory," Carl Sagan in *Cosmos* stated that humans are made of "starstuff."[28] The starstuff is chemicals that appear in DNA, which is the combination of chemicals that determine our development as living beings. But what is the process from stars to chemical elements in DNA for human life?

In Sagan's account the universe is 15-20 billion years old.[29] (A more recent estimate is 13.7 billion.) The Big Bang at the beginning of the universe produced both hydrogen and stars. (Stars supposedly came about from concentrations of cosmic dust.) Hydrogen is the simplest of all chemical elements with one electron going around one proton. Apparently following views first expressed by the British astronomer Fred Hoyle, Sagan believed that stars took hydrogen from the Big Bang and cooked it into heavier atoms.[30]

According to Sagan, after Earth formed 4.6 billion years ago, atoms of various kinds from the stars arrived on our planet. (The belief is that some stars exploded and sent various materials to Earth.) Chemicals made from these atoms provided for the beginning of life 4 billion years ago.[31] Rays from the sun supposedly started life in a kind of chemical soup.

Sagan thought that one-celled plants appeared 3 billion years ago.[32] Sex arose about 2 billion years ago.[33] The Cambrian explosion of life forms occurred 600 million years ago.[34] Trilobites (tiny sea animals about the size of insects) came into existence about 500 million years ago but ended about 200 million years ago.[35] Less than 10 million years ago there were creatures resembling humans, who were followed by humans a few million years ago.[36] (Sagan's idea of the first humans suggests that monkeys or apes came down from trees and started walking on two feet.)

The DNA (deoxyribonucleic acid) part of this theory emphasizes the chemicals that give instructions for the development of life forms. All forms of life on our planet today have some kind of DNA.

Sagan tells us the human body has a hundred trillion cells.[37] Each cell has a nucleus. The nucleus contains DNA, which is a series of different combinations of four chemicals paired like rungs in a twisted ladder. The chemicals in the double helix of DNA determine how any living body will develop. In the view of Sagan and many others, DNA helps to show the origin of life in chemicals produced by stars.

In contrast to the published views expressed by Darwin, there is no mention of the Creator in the stars-DNA interpretation of life by Carl Sagan. There are instead purely scientific explanations with no religious references.

Yet another scientific interpretation of life is, with no disrespect intended, the ignorance theory. Darwin himself claimed there was much ignorance on the origin of species with much not explained.[38] Darwin was not the only person with reservations and admission of lack of knowledge.

Tyson and Goldsmith indicate there is knowledge of approximately when life began but not knowledge of where or how.[39] Without giving supporting facts, they assert there is definite evidence of life on Earth 2.7 billion years ago.[40] They wrote that most paleobiologists believe that life on our planet began at least 3 billion years ago.[41] Tyson and Goldsmith think life may have begun, may have been wholly or mostly exterminated, and then may have begun again, with this process occurring more than once.[42] They also consider the view that the origin of life may be related to extremophiles, beings that live under extreme conditions such as high temperature, acidity, or high pressure.[43] In spite of various possibilities, Tyson and Goldsmith claim the origin of life on Earth "remains locked in murky uncertainty."[44]

Religious interpretations of life usually express conviction that God is responsible for life. As to what is known about life, especially its origins, scientific interpretations vary from the highly informed but humble approach of Darwin to the confidence of Sagan to the open but hesitant outlook of Tyson and Goldsmith. Scientific interpretations of life seldom refer to God.

Analysis

The differences between religious interpretations of life and scientific interpretations of life are strong and sometimes lead to hostility. What, if anything, can be done to improve the situation?

Scientific interpreters of life could do a better job of explaining their views. Comparatively few people are going to read Darwin's books and other difficult works. Even if people are willing to read, scientific writings are usually not easy to understand. The general public needs additional help. Scientists or, at least, scientific educators should try to explain in a non-technical way not only the basic views of scientists but also the evidence they have.

What should scientists try to explain? What is the scientific evidence for when and how and why life began? What is the scientific evidence for when and how and why animal life became separate from plant life? What is the scientific evidence for when and how and why different sexes or genders occurred? What is the scientific evidence for when and how and why human life began? What is the best scientific evidence for the sequence in which major life forms appeared?

Scientific interpreters of life should also be reminded occasionally that much of what is claimed to be scientific knowledge goes against common sense. When it comes to thinking that new and different life forms have somehow come about from much older life forms, common sense may lead us to be skeptical. We have not seen such changes in our lifetime, and we do not know of any people in the historical past who have seen such changes. The amount of time required for these claimed changes is astounding. Evolutionary biologists should understand that common sense is often against their ideas even apart from any religious considerations. Reality may go beyond common sense, but common sense is the place most of us begin.

Religious interpreters of life should recognize, for their part, that even common sense supports biological evolution to some extent. If we think of biological evolution as being gradual change over time, how much change and how much time might we consider?

We know that individual humans change gradually but significantly from infancy to childhood to adulthood to old age. We are considering a period of eighty to ninety years. It is often hard to recognize a senior adult from a baby picture. Even more drastic is the change that occurs in humans from conception to birth. It usually takes nine months to go from what is very small and unformed in the womb to the birth of a baby. The beginning of a new life at conception looks nothing like what develops by the time of birth. All of us can recognize biological evolution from conception to birth and from infancy to old age in the lives of individuals.

It is reasonable also to think that many humans today may be taller and stronger and healthier than the average person of several centuries ago. There have been advances in medicine and nutrition that could help bring about such changes. It is not difficult to think that humans may have changed for the better biologically over the course of many generations. If we are thinking of this much change (better bodies) over this much time (possibly centuries), most people could accept this kind of evolution.

Many people have difficulty with the claim that a great amount of change occurred in kinds of life forms over an extremely long period of time. Individuals can experience great changes within their lifetimes. Groups of individuals of a certain kind can change in some ways over centuries from earlier groups of individuals of that kind. But can one species or life form eventually lead to a new and different species? Was Darwin correct that a process beginning with marine animals could, over millions or billions of years, bring about mammals? Do humans have ape-like beings as distant ancestors? However far common sense may lead us in accepting some change in life forms, the claims by Darwin and others go well beyond common sense.

Whatever difficulties Darwin's views may present for common sense, is it possible to combine religious and scientific interpretations of life? In their book mentioned earlier, Ted Peters and Martinez Hewlett ask this question in their title *Can You Believe in God and Evolution?* Many people would answer in the negative, but some have answered in a positive way.

Darwin obviously believed in his own views of evolution. In addition, he referred to the Creator in *The Origin of Species*. In a passage already cited, Darwin wrote that many wonderful forms of life have come from the few forms or one originally breathed by the Creator.[45] In a passage in *The Descent of Man* also previously mentioned, Darwin wrote that the question of the existence of a creator had been answered affirmatively by some of the world's highest intellects.[46] Whatever Darwin personally concluded about God, his published statements allow for acceptance of both a creator and evolution.

Pierre Teilhard de Chardin, a Roman Catholic paleontologist, is well known for trying to combine religious and scientific interpretations of life. He thought that God not only was responsible for the beginning of life but also actively guided the evolutionary development of life.

A recent advocate of belief in both God and evolution has been Dr. Francis Collins, a physician and also director of the Human Genome Project. Dr. Collins did report some disagreement and disappointment with his views when he presented himself as a theistic evolutionist at a national meeting of Christian physicians.[47]

Ted Peters and Martinez Hewlett themselves have a special interpretation of God and evolution in relation to God's goal in evolution. They suggest interpreting Genesis 1.1-2.4a in the light of Revelation 21.1-4. Their idea is to think of Creation in relationship to New Creation.[48] The belief appears to be that God is directing the evolution of Creation to the goal of New Creation.

While there are various ways for believing both in God and Darwinian evolution, it is not possible to believe both in Darwinian evolution and a literal interpretation of the first two chapters of Genesis. There are conflicting accounts by the Bible and Darwin.

According to Genesis 1 and 2, God separately created plants, sea creatures, birds, animals, and humans. The creation of humans occurred directly without any development from previously existing forms of life.

As to Darwin's views, he admitted he was not sure whether the Creator breathed one or many forms of life at the beginning. When Darwin turned to the genealogy of man, he began with tiny sea creatures. The sea creatures eventually brought about fish and amphibians. Some of the fish developed into mammals. Birds and reptiles also came into existence from fish. The mammals included ape-like animals or Old World monkeys from which humans separated. Darwin's view has humans appearing long after the beginnings of other forms of life and does not directly connect God with humans.

There appears to be no way that a literal interpretation of all of Genesis 1 and 2 can be reconciled with Darwinian evolution. For those who claim to obtain their truth, including scientific truth, from a literal interpretation of the Bible, there is no choice but to disagree with Darwinian evolution. From the viewpoint of biologists who follow Darwin, there cannot be acceptance of Genesis 1 and 2 as literally true. It is difficult to see any way in which the opposing interpreters could come to an agreement. Is there any possibility of a change?

Those who have a literal view of the first two chapters of Genesis have a problem with biblical interpretation, namely that Genesis 1 and 2 have two differing stories of creation. The creation account in 1.1-2.4a differs in several ways from that of 2.4b-25. Since there are conflicting differences in the two creation accounts, they cannot both be literally true.

One difference in the two creation accounts is the time frame. In Genesis 1 there are six days of creation with rest by God on the seventh day. Humans were created on the sixth day. In Genesis 2 there is only one day of creation. We read in 2.4 about "the day that the Lord God made the earth and the heavens." According to 2.7, it was also at that time that "the Lord God formed man from the dust of the ground." This second account has God's creating man the same day that God made the earth and the heavens. Without any indication of a change in time, the story then covers God's planting a garden in Eden, forming animals and birds as possible helpers for man, and finally creating woman. How could there literally be both seven days of creation (with a day of rest) and only one day of creation?

A second difference in the two creation accounts is the order of creation. According to Genesis 1, God created vegetation on the third day, brought forth creatures of the sea and birds on the fifth day, and on the sixth day created animals and then humans. Genesis 1 has humans as the last living things God created. When we turn to Genesis 2 we find a different order. God created man before planting a garden in Eden and making trees grow out of the ground. God created man before creating animals and birds as possible helpers for man. If Genesis 1 has animals created before man but Genesis 2 has animals created after man, how could both accounts be literally true?

A third difference in the two creation accounts is a contrasting view of the female. According to Genesis 1, God created male and female at the same time. There is no indication of creation of one gender before the other. Both male and female were created in the image of God, and both were commanded to be fruitful and multiply and to have dominion. The implication is that male and female were created equal.

According to Genesis 2, God first created man and then animals and birds as possible companions for man. When animals and birds were insufficient for man, God created woman. Since God created the woman out of the man's rib, the man recognized the woman as part of himself. There is a closeness between male and female, but man came first. The woman was made because God thought the man should not be alone and because other creatures were not adequate companions. If Genesis 1 has male and female created at the same time or at least with no distinction in time but Genesis 2 has the woman created not only after man but also after other beings, how could both creation accounts be literally true?

The proper interpretation of Genesis 1 and 2 may be debated. Is Genesis 2 a more detailed account of what is covered in Genesis 1? This interpretation is unlikely because of the significant differences in the two accounts. If Genesis 1 and 2 are not both literally true, should we choose one over the other? Should we take the view that both accounts should not be interpreted literally? Do the accounts provide a religious interpretation of life with God as creator but lack a scientific understanding of how various life forms began and developed?

Religious interpreters of life sometimes look beyond the Bible to find support for their views. They would like to find evidence that there is indeed a creator behind life, evidence that scientific interpreters of life would have to accept. An example of proposed evidence is the eye, an amazing organ of great complexity. It is the chief means for making observations, which are the basis for science. With all of its intricate and wondrous operations, does not the eye point to a great designer whose power and intelligence are absolutely necessary for its existence? It does not seem possible that such a complicated and amazing organ could have evolved. Does not the ingenious working of all the parts of the eye point to a creator?

Scientific interpreters of life are not convinced there has to be an explanation for the eye beyond evolution. Darwin himself had a section in *The Origin of Species* on organs of extreme complication. He admitted that it seemed absurd to think the eye could have developed by natural selection. He still believed that this "living optical instrument" came about over millions of years.[49] Darwin referred to the Creator but did not believe God was directly responsible for the eye.

Much about interpretations of life has to do with attitudes toward the Bible. Is the Bible completely correct about everything it covers? Is the Bible inspiring in many ways but scientifically outdated? If the Bible is not always taken literally, could many of its views be reconciled with science?

Conclusion

Life is extremely valuable for almost all beings but is hard to define with precision. Characteristics of life are the consumption of nourishment, the process of growth, and the ability to reproduce. Hardly anyone would think of those qualities as a sufficient description of the nature and value of life.

Religious and scientific interpretations of life differ. A common religious interpretation of life, based on literal interpretation of the Bible, includes belief in the immutability of species as originally created by God. Darwin disagreed with belief in the immutability of species. He claimed major changes in life forms, including changes from one species to another, over vast periods of time. However, Darwin referred in his writings to the Creator as the source of life, whereas most modern scientists do not think science leads to belief in a creator and try to explain life only in natural processes.

Finding common ground is difficult. Many religious and scientific interpreters do not wish to change their views. One possibility for change is to have a religious interpretation of life that accepts belief in God as creator, does not take all of Genesis 1 and 2 literally, and allows for scientific views of evolution.

Questions for Further Thought

1. How would you define the nature of life?
2. Should Genesis 1 and 2 be interpreted literally? Why or why not?
3. What scientific evidence, if any, do you think points to a creator of life?
4. Do you think both Darwinian evolution and belief in a creator should be taught in the science classes of public schools? Why or why not?

Notes

[1] Donald Goldsmith and Neil deGrasse Tyson, *Origins* (New York: W. W. Norton & Co., 2004), 234.
[2] Charles Darwin, *The Origin of Species* (1879; reprint Edison, NJ: Castle Books, 2004), 51, 66.
[3] See the diagram in Darwin, *Origin*, 140-41.
[4] Darwin, *Origin*, 149.
[5] Ibid., 155.
[6] Ibid., 655.
[7] Ibid., 617.
[8] Ibid., 662-63.
[9] See especially ch. 10 in Darwin, *Origin*.
[10] Darwin, *Origin*, 156.
[11] Ibid., 658.
[12] Charles Darwin, *The Descent of Man* (2nd ed., John Murray, 1879), 22.
[13] Ibid., 25.
[14] Ibid., 28.
[15] For summaries of evidence, see Darwin, *Descent*, 42-43, 675-76.
[16] See Darwin, *Descent*, 172-82.
[17] Ibid., 182.
[18] Ibid., 183.
[19] Ibid., 184.
[20] Ibid., 185.
[21] Ibid., 192-93.
[22] Ibid., 689.
[23] Ibid., 193.
[24] Ibid., 669-70.
[25] Ibid., 116.

[26] Richard Dawkins, *The Greatest Show on Earth: The Evidence for Evolution* (New York: Free Press, 2010), 403-404. For further information on what Darwin may or may not have believed and when, see, for example, Keith Thomson's Introduction in *The Religion and Science Debate: Why Does It Continue?*, ed. Harold Attridge (New Haven, CT: Yale University Press, 2009), 10-11.

[27] See Ted Peters and Martinez Hewlett, *Can You Believe in God and Evolution?* (Nashville: Abingdon Press, 2008), 25-29.

[28] Carl Sagan, *Cosmos* (New York: Ballantine Books, 1985), 12.

[29] Ibid.

[30] Ibid., 179. For a statement that all chemical elements beyond helium were forged in stars, see Tyson and Goldsmith, *Origins*, 159.

[31] Sagan, *Cosmos*, 20.

[32] Ibid., 21.

[33] Ibid.

[34] Ibid., 22.

[35] Ibid., 23.

[36] Ibid.

[37] Ibid., 21.

[38] Darwin, *Origin*, 156.

[39] Tyson and Goldsmith, *Origins*, 248.

[40] Ibid., 236.

[41] Ibid., 121.

[42] Ibid., 239.

[43] Ibid., 244-48.

[44] Ibid., 235.

[45] Darwin, *Origin*, 669-70.

[46] Darwin, *Descent*, 116.

[47] Francis Collins, *The Language of God* (New York: Free Press, 2006), 146.

[48] Peters and Hewlett, *Can You Believe?*, 84-85.

[49] Darwin, *Origin*, 223-28.

CHAPTER 10

God and Space

Thinking about God and science has led us to consider the important subjects of time and life. Space is another significant concept, especially the extent of space. How should we think of God in relationship to space? The relationship between God and space and humans is especially challenging.

When we think of the nature of space, there are two basic and opposing ideas, namely that space is nothing (empty space), and space is whatever is filled by something (occupied space). When we think of the extent of occupied space and how to measure it, people have gone from very small units to enormous distances.

Religious interpreters have tried to emphasize humans as central in space and as significant to God. Scientific interpreters have emphasized the vastness of space, often without special regard for God or God's relationship to humans. The scientific view of the staggering size of the universe in space presents difficulties for religious interpretations of God and humans.

The Nature of Space

Space, like time and life, is something we all think we know but that can be difficult to describe. In the view of space as nothingness, we cannot correctly say what space is because it is not anything. Space in this understanding is the absence of everything. We may speak of empty space, which is an area occupied by nothing. We thus have the "emptiness" understanding of space. If an area of space is not occupied by anything, then there is a vacuum. It is difficult and perhaps impossible on our planet to arrange a complete vacuum, but we can imagine the possibility. Space then would be an empty area that has nothing in it. Since we have difficulty in trying to think of nothing at all, we usually think of empty space as having something around it.

An opposing view of space describes it as having something in it. In this view, we then have the filled or occupied version of space. Something "takes up" empty space by replacing nothing with something. Even space that does not appear to have anything in it might have something such as air. Space in this view is not exactly an ocean but has a similarity. Space would, somewhat like an ocean, include all existing objects within itself. What would all of occupied space look like? We might think of a ball or globe, but perhaps it would be more like a football or watermelon. Whatever all of occupied space might look like, we can think of additional unoccupied space beyond occupied space.

If we try to put the two concepts together in relation to the universe, there can be some confusion. Are there areas of our universe that are empty space where there is nothing at all? If so, then how do empty space and occupied space maintain distinct areas? Is it possible that areas of our universe that appear to be empty actually contain something? What are we supposed to think about dark matter and dark energy? Does the known universe have areas of completely empty space, or is all of it occupied by something—including things we cannot see?

Then there is the question of what is beyond our universe of at least largely occupied space. Is there occupied space beyond our universe that scientists have not yet discovered? If so, how much additional occupied space is there? Does space exist beyond our universe? If so, how much empty space is there? Does it even make sense to ask how much nothing (unoccupied space) there may be beyond our universe of occupied space? It is easy to see how we can become confused when trying to think about the nature of space for the entire universe and beyond it.

The Extent of Space

If we deal only with space as emptiness, there would be no way to measure it and determine its extent. How could you measure nothingness by itself? In order to consider the extent of space, we have to have something to measure. Only occupied space has something to measure. What we measure is not space as such. We measure objects in space and the distances between them.

One way of measuring space (occupied space) is to use parts of the body. The width of a man's thumb is the basis for an inch. The width of a man's hand has been used as a unit to measure, among other things, the height of horses. A horse would be a certain number of hands high. A man's foot has also been used as a unit of measurement. A unit probably used in the building of the Egyptian pyramids was a cubit, which was the length of a man's forearm from the elbow to the tip of the longest finger. We might wonder which man was used for these units or if there was an average of various men.

The old English system of measurement, still followed in the United States, used such units as inches and feet. Once the basic units were established, people could be measured from head to toe in feet and inches. When longer distances were calculated, yards and rods and miles were used, all of which could be expressed in feet.

When earlier people thought of even longer distances, they probably were not thinking of much. A few hundred miles across land or sea would have seemed enormous to them. They could see the sun, the moon, and the stars without knowing much about distances from our planet.

Today there is astounding knowledge of the extent of space. Consider the distance from our planet to the sun. Earth's orbit has an average distance of 93 million miles from the sun.[1] We might think the distance to the sun is impressive, but there is far greater distance from the sun to the edge of our solar system. Then there is much beyond our solar system.

Before we think more about distances, let us consider the size of objects that are to be in our calculations. While our solar system is extremely large, it is a comparatively small part of the collection of stars that form our galaxy. The Milky Way galaxy, our galaxy, has about 400 billion stars.[2] The Milky Way is quite large, but how many galaxies are there? There are a hundred billion galaxies with an average of a hundred billion stars per galaxy.[3]

Feet and yards and miles are not the best measures for distances between objects in deep space. There has been use of a light-year. That term designates the distance light would travel in one orbit of our planet around the sun. Since light travels at approximately 186,000 miles per second, we encounter some big numbers. A light-year is approximately 5.878 trillion miles.

When distances in space are expressed in light-years, there are prodigious numbers. The distance from our sun to the center of the Milky Way is 30,000 light-years.[4] That distance would be 30,000 times 5.878 trillion miles. The Milky Way galaxy is about a hundred thousand light-years across.[5] How far is it, through billions of other galaxies, to the edge of the known universe? Calculations can change, but one measure of the distance from Earth to the most remote quasars at the edge of the universe is eight or ten billion light-years.[6] With most of us, our eyes begin to glaze over, if they have not already done so.

If these numbers are not big enough, consider another thought. The mammoth universe of which we are a part is expanding. It was Edwin Hubble who discovered evidence in the 1920s that the universe was rapidly expanding.[7] The steady state theory of the universe held by Fred Hoyle and some others was abandoned.[8] While the universe as a whole is expanding, the parts of the universe farthest from us are accelerating more rapidly than the parts of the universe closest to us.

We might wonder, from a scientific point of view, how much longer the universe will expand and how huge it will become. Will the universe keep expanding forever? Will there perhaps be a reversal and a decrease in size, perhaps a Big Crunch, in ten billion years or so? At present, the increase is occurring.

If the universe is expanding, what allows this activity? The universe must be moving into empty space or at least into space less densely occupied than what is in the universe. What kind of space (empty or occupied) is beyond our universe? How far does this space—whatever it is—go? Is the space beyond our universe limited or infinite?

We can imagine something, even if it is only empty space, beyond the space occupied by our universe. We have a difficult time trying to imagine how far occupied space can expand. The distance must be colossal.

God's Place

What are the implications of the vast extent of space for beliefs about God? What is God's place in relation to space? Answers vary with religious and scientific interpreters.

Let us start with a religious interpretation of space. According to the first chapters of Genesis and other parts of the Bible, God created the heavens and the earth. With the view that the Creator is at least in some ways greater than the Creation, God is great indeed. If the extent of the universe is amazingly big, the one who provided for it must be considered with high respect and awe. If we are staggered by the size of the universe, we should be overwhelmed with admiration and reverence for the Creator.

But what if the creation accounts in Genesis are not literally true? What if there was a Big Bang instead of the six days of creation in Genesis 1? For all religious interpreters, God was still the Creator but perhaps used the Big Bang.[9]

What about the views of scientists? Stephen Hawking rarely refers to God in *A Brief History of Time,* and even then there is the question of how serious he is. For example, Hawking mentioned the possibility that the universe is an act of God to create beings like us.[10] But Hawking also wrote of the possibility that the universe is self-contained. If so, what place is there for a creator?[11] Hawking appears to be not much impressed with the greatness of God or with God at all.

If we turn to scientists such as Carl Sagan and Neil deGrasse Tyson, they do not even mention God—at least not in their works considered here. These men have a different interest in relation to the vast expanse of space in our universe. Is there intelligent life anywhere other than on our own planet? For Sagan and Tyson, God appears to be an outdated belief. The exciting prospect for them is that there may be some form of intelligent life out there in the vast reaches of space.

If we return to the belief that God and space are both real, there are still some difficulties. With the view of traditional theism that God is omnipresent, is God present in both occupied space and unoccupied space? According to the traditional view that God is in heaven, is heaven a reality within space or somehow above and beyond space? What should we make of the idea in traditional theism that God is both immanent (within the world) and transcendent (above and beyond the world)? Is God within space as well as above and beyond space? If God is above and beyond space, is God in a different dimension we cannot comprehend except perhaps in a figurative way?

At least some religious interpreters may understand the vast extent of space as a challenge to think more deeply about the greatness of God. Some may recall the title of the book by J. B. Phillips, *Your God Is Too Small*. Should the greatness of space indicate for believers that God is far greater than even some devoted followers may have thought?

There is, however, a widening gap between religious and scientific interpretations of space and God. Religious interpreters can ponder the vastness of occupied space and think how great God is. Scientific interpreters of occupied space are tending more and more to omit any consideration of God.

Our Place in Space

The vast expanse of occupied space in our universe gives us reason to wonder about the significance of our place in space. We humans are beings who do not live very long, usually less than a hundred years. We inhabit a small planet, much smaller than some other planets in our solar system. Our solar system is itself a small part of the immense Milky Way galaxy, and there are billions of other galaxies with many billions of stars. How significant are we in the universe? As we learn more about the universe, our significance appears to decrease. If we have little significance in the universe, how significant are we to God?

Long ago, people thought the world was flat and that the sun, at whatever distance, circled a stationary earth. Ptolemy, an Alexandrian astronomer of the second century A.D., developed a system in which Earth was the center of the universe. Almost everyone had already accepted a geocentric or earth-centered understanding of the universe. It was very easy then for people to think of themselves as significant.

Not everyone agreed with the earth-centered view of the universe. Aristarchus of Samos, who was active in the third century B.C., may have been the first to hold a heliocentric or sun-centered view of the universe.[12] There is a story that Archimedes chided Aristarchus for having a nonsensical idea.

The heliocentric theory of the universe was again advocated in the late Middle Ages. Nicolaus Copernicus (1473-1543) and Galileo Galilei (1564-1642) both accepted the view of a sun-centered universe. Their idea was that not only Earth but all of the planets in our solar system revolved around the sun. Galileo offered use of his telescope to provide evidence for his conclusion.

Many church officials of the time were alarmed by Galileo's ideas and refused to look through his telescope. What was their problem? The heliocentric view of the universe did not change the distance from our planet to the sun. But the switch from our world as the center of the universe to the sun as the center had religious implications. If Earth was not the center of the solar system, how important were its people to God? Some might give the dangerous answer "not very." Many churchmen of the time claimed that Galileo was going against the Bible.[13] The Inquisition condemned Galileo for heresy, and he was forced to recant his views.

Officials of the Roman Catholic Church have had additional thoughts about the decision against Galileo. There have been adjustments through the centuries with more favorable views of Galileo and his beliefs.

In 1992 Church officials spoke of good faith by all parties in Galileo's case. They also referred to the remarks of Cardinal Robert Bellarmine in 1615 that explanations of biblical texts should not conflict with established truth. At the time, people thought the earth-centered view of the universe was established truth. After other scientists overwhelmingly showed that the sun-centered view of the universe was correct, officials of the Catholic Church eventually recognized the need for some revision concerning Galileo. Pope John Paul II said in 1992 that theologians of Galileo's time had commendable pastoral concern for church teaching. He added, however, that they did not distinguish carefully enough between teachings of sacred scripture and their own interpretations.[14]

From the viewpoint of both modern scientists and Galileo, where are we in the universe? According to the heliocentric interpretation, our place in space is not at the center of our solar system. The sun is at the center and is orbited by planets of various sizes, our planet being one of the smaller bodies. In addition, our sun is not at the center of the Milky Way galaxy.[15] If we are not at the center of our galaxy, where are we? Our galaxy is a spiral galaxy, and we are near the edge of a spiral arm of our galaxy.[16] Where is our galaxy in relation to the universe as a whole? Hawking claimed not to know if our galaxy is at the center of the universe.[17] Tyson and Goldsmith have stated we are not at the center of our solar system or our galaxy or the universe.[18]

Where then is our place in space? It may be understandable that some of us are confused about exactly where we are in the universe. People have gone from thinking they were at the center of the universe to being told their planet is not the center of anything. How can we claim any significance in the universe? If our whole planet somehow disappeared, would there be much effect on everything else that occupies the vast expanse of space? From the scientific point of view, it is difficult to find much significance for humans in the vastness of occupied space.

For religious people, there is the question of our significance for God. If our universe is exceedingly large and God is greater than the universe, why should we think we have any special significance for God? Why should God pay any attention to beings with short lives who live on a small planet in a solar system that is itself only an extremely small part of one of billions of galaxies?

Assurances of Our Significance

Although our place as humans in the vastness of space may look extremely small and unimportant, we can think of ways to assure us of our significance. These ways may not be convincing to everyone, but we are not hopeless. We can think that we do have significance in the universe and also in God's sight.

One assurance of our significance, at least for religious people, is the Bible. David expressed concern about our possible lack of importance to God in Psalm 8.3-4: "When I look at your heavens, the work of your fingers, the moon and the stars that you have established; what are human beings that you are mindful of them, mortals that you care for them?" An answer immediately follows in verse 5: "Yet you have made them a little lower than God, and crowned them with glory and honor." David was deeply impressed with the grandeur and immensity of the universe he could see. Yet he was still convinced that people are important to God.

We also read in the Bible that God, after creating the heavens and the earth, created people. He gave commands for people to follow. God also provided means of forgiveness when people failed to follow the commands.

We read that God is holy but also good, including being good to people. We read that God is perfect but also is love. We read in the New Testament that God came in human form and that Jesus, the Son of God, died to save all who would believe. There are many indications in the Bible of how humans have often failed to serve God properly, but there are also numerous teachings about God's concern for people.

It is true that the worldview of biblical writers is geocentric rather than heliocentric, earth-centered rather than sun-centered. We do not read in the Bible about planets in a solar system or about galaxies or black holes in space or quasars. The men who wrote the books of the Bible did not know of the discoveries that would be made by scientists. Perhaps the Bible would be more reassuring if there had been some expression of God's concern for humans in spite of the vast number and immense size of galaxies. Religious interpreters remain convinced that the Bible reveals the importance of people to God.

Another assurance of our significance is the testimony of many that God is present and active in their lives. There is the view that God has, at least sometimes, provided healing that might not otherwise have occurred. Many believe that God has helped them in their daily lives either by preventing something bad from happening or providing something good or both. There is the conviction that prayers have been answered, not always but when the prayers were in accord with God's will. Many find assurance in their own experiences beyond the Bible that God cares for them.

An assurance of our significance apart from the Bible or religion is that even comparatively small things can still be important in some way. Humans are much smaller than gigantic objects in space but can still be significant. Consider the thought that even objects much smaller than humans have some significance. Think of living organisms in our water that can be seen only through a microscope but that can make us ill. Has your health ever been affected by what doctors tell us is a tiny virus that your eyes do not detect? Have you ever seen iron in your blood that improves your strength? Very small things can significantly influence our lives for good or ill.

A special example of the significance even of very small things is the atom. Democritus and other ancient Greeks thought an atom was the smallest of all objects. The name "atom" refers to what cannot be cut or divided, that is, made any smaller. Democritus did not anticipate things smaller than an atom. He did not know there would be the discovery of subatomic particles. Scientists now tell us not only about an atomic nucleus of protons and neutrons circled by electrons but also about extremely small quarks. There are supposedly three quarks within each proton and neutron. And perhaps smaller than quarks may be virtually infinitesimal "strings" of electrical impulses. Atoms are small things made up of even smaller things. But atoms combine into molecules to produce objects that make up the universe.

Also, scientists discovered how to split atoms and create atomic energy. Einstein expressed the importance of some small things when he gave his famous formula for atomic energy. He said that energy equals the multiplication of mass by the square of the speed of light. Since light travels at approximately 186,000 miles per second, the square of the speed of light is an impressive number. The idea was that a very small amount of material can, under certain conditions, produce a great amount of energy

We can see from everyday life and from science that a relatively small size does not necessarily keep something from being significant. The small size of humans in the universe is not the only consideration in judging their importance. But is there something special about humans that, despite their comparative smallness, would indicate their value?

One more assurance of the significance of humans comes from a remarkable quality. Among other abilities, humans can think. Was it not Pascal who said that man is a reed but is a thinking reed? As with a reed that is easily bent or broken, it is not difficult to kill a man. But a man is greater than the universe itself in one way: Man can think while the universe cannot.

Who is concerned about space, including great objects and huge distances in outer space? Dolphins and chimpanzees and eagles appear to have no concept of the solar system and galaxies and quasars. Galaxies, as immense as they are, give no indication of having any kind of intelligence. Humans are comparatively small in the universe, but we all have the capacity to think at some level. The ability of humans to think does not guarantee correct answers to every question, but who or what else is even asking questions about space? This ability to think is an important indication of the significance of humans in the vastness of the universe.

Further, who is it that considers God? We know of no animals that have such an idea. There is no proof of any other intelligent life in the universe. The ability to think about God is itself a special indication of the significance of humans.

What is the source or explanation for thinking? Should it be understood as an evolutionary development over a vast amount of time? That view appears to be the prevailing scientific interpretation. Many people, however, are convinced that the thinking ability of humans is not an accident; it is a gift from God that helps show our significance for the universe and for God.

Conclusion

When we consider the nature of space, we think of empty space (unoccupied) and filled space (occupied). If we turn our attention to deep space or outer space, there appear to be huge areas of unoccupied space. It may be that apparently unoccupied areas of space are actually occupied by such things as dark matter and dark energy.

If we try to think of the extent of occupied space in terms of the size of the universe, we encounter truly astronomical numbers. It is 93 million miles from our planet to the sun, but there are even much bigger numbers. Our entire solar system is a small part of the Milky Way galaxy, which is tens of thousands of light-years across. (A light-year is the distance light travels at approximately 186,000 miles per second in one orbit of Earth around the sun.) There are many other galaxies in the universe. At the edge of the known universe are the quasars. The distance from Earth to the quasars, the known objects farthest from us, might be 8 or 10 billion light-years, perhaps more.

When discussing space, many scientists say little, if anything, about God. They are often more interested in whether or not there is other intelligent life in the universe.

How does the extent of space, the gigantic size of the universe, affect religious views? One interpretation is that the immense size of the universe should help us to understand that God as creator is far greater even than the vastness of space.

Another religious view seeks to establish the significance of humans in view of a possibly discouraging question. If God is greater than the staggeringly immense universe, is God concerned about humans with our short lives on a small planet in one of perhaps a hundred billion galaxies?

The prevailing view of our place in space has changed over the centuries. People once thought the sun circled our planet and we were the center of the universe. It was not too difficult to think that God paid special attention to humans. With the heliocentric view, the sun is the center of our solar system. Even more, scientists today claim our solar system is not the center of our galaxy and our galaxy is not at the center of the universe. Because of various scientific claims, it is more difficult now than earlier in history to think we are important to God.

In spite of the vast size of the universe and our much-less-than-prominent place in it, there are some assurances of our significance both in the universe and in God's sight. The Bible has many teachings about God's interests in humans from his creating the first people to sending his Son to die for the sins of people. Also, many claim that God is present and active in their lives.

If we go beyond religious considerations, there is the idea from everyday life and from science that even comparatively small things do have significance. There are tiny beings or substances we cannot see but that affect our health. Also, although atoms are so small we cannot see them, scientists can unlock tremendous energy from them.

If we wonder about any special quality that allows humans to be significant, there is thinking. No other beings in the universe, as far as we know, have the capacity for thinking that humans have. Religious interpreters believe that thinking is a gift from God. Our thinking helps assure our significance to God in spite of God's greatness and our tiny place in the vastness of space.

Questions for Further Thought

1. Why do you think we read references in the Bible to the sun and the stars but see no mention of planets and galaxies?
2. When you consider the scientific view that the universe is huge and is expanding, what do you think is beyond the universe?
3. In view of the vast extent of space, should we change our view of God? Why or why not?
4. In view of the vast extent of space, how would you explain the importance of persons to God?

Notes

[1] See, for example, Neil deGrasse Tyson and Donald Goldsmith, *Origins* (New York: W. W. Norton & Co., 2004), 267.

[2] See Carl Sagan, *Cosmos* (New York: Ballantine Books, 1985), 4. See also Tyson and Goldsmith, *Origins*, 27.

[3] See Sagan, *Cosmos*, 3.

[4] Ibid., 165.

[5] See Stephen Hawking, *A Brief History of Time*, 10th ed. (New York: Bantam Books, 2001), 39.

[6] See Sagan, *Cosmos*, 165.

[7] See Hawking, *Brief History*, 41.

[8] Ibid., 49-50.

[9] Stephen Hawking has noted that the Catholic Church in 1951 officially accepted the Big Bang model as being in accord with the Bible. See *Brief History*, 49.

[10] Hawking, *Brief History*, 131.

[11] Ibid., 146.

[12] See Sagan, *Cosmos*, 12 and 155. See also Tyson and Goldsmith, *Origins*, 204.

[13] See the account in Joshua 10:12-14 of how the sun, usually thought of as moving, stood still for about a day.
[14] See *L'OSSERVATORE ROMANO,* English ed., 4 November 1992.
[15] See Hawking, *Brief History,* 39. See also Sagan, *Cosmos,* 158.
[16] See Sagan, *Cosmos,* 3.
[17] Hawking, *Brief History,* 44-45.
[18] Tyson and Goldsmith, *Origins,* 230.

AUTHOR'S VIEWS

As we have tried to think carefully about God, we have considered God and the Bible, God and philosophy, and God and science. Some of my views have probably been apparent, but I will try to be more specific.

In our discussion about God and the Bible, it was evident the Bible does not have a comprehensive summary of beliefs about God. Rather, it contains various stories and scattered statements about God. The approach here has been to use those elements in proposing a summary of biblical teachings about God and to consider two areas of special concern: (1) God and violence and (2) God and human destiny.

I appreciate and accept the biblical teachings about God that we have covered under the headings of defining qualities, descriptive titles, and divine actions.

As to defining qualities, various "God is" verses in the Bible refer to God as one, good, holy, and perfect. Additional verses state that God is Spirit, light, and love. All of these teachings are profound,

Some biblical teachings portray God with descriptive titles such as creator, judge, redeemer, shepherd, king, husband, and father. All of these titles have important meanings, but some of them should be understood symbolically. When we think of God as husband, for example, the idea is not literally someone who has gone through a marriage ceremony but one who expects and provides faithfulness. We should not let literal interpretations obscure spiritual meanings.

When we considered actions by God in the Bible, we saw that the Old Testament has references to creating, commanding, punishing, saving, making covenants, and calling. The New Testament writers refer to such divine actions as incarnating, raising Jesus from the dead, sending the Holy Spirit, revealing, reconciling, and working for good. All of these actions are impressive.

One area of special concern regarding God and the Bible is violence. There are many reports of God and violence in the Bible. Since violence is force that usually results in harm of some kind, we might wonder how a loving God could be associated with harm.

We can find three important explanations in the Bible about God and violence. The main explanation in the Old Testament is the punishing action of God. Although God is loving, God hates unrighteousness. As a judge, God punishes evildoers. We also find in the Old Testament the explanation that God allowed violence against Job as a test of Job's faith. God permitted Satan to bring harm into Job's life in order to show that Job would remain faithful. In the New Testament, we read indications that God was ultimately behind the violent death of Jesus. Various references in the New Testament suggest the explanation of a redemptive purpose by God for Jesus' death.

I am glad the Bible has some explanations for God and violence, even if the explanations are sometimes indirect. I do wonder about the severity and extent of violence by God reported in the Old Testament. The accounts can be unsettling. Also, I am disappointed that the story of Job's faithfulness despite violence does not give more credit to Job. I wish God had given an explanation to Job and had told Job that he passed the test. In the case of the crucifixion of Jesus, God's expression of love through a violent act sounds jarring but is profound. The voluntary suffering of an innocent person for others does not sound fair but is admirable.

A second area of special concern regarding God and the Bible is human destiny. The Bible has many references to human destiny, but those references leave much unanswered and often do not have a direct reference to God. I do not find in the Bible a comprehensive view of human destiny that is completely clear and coherent. Much depends on interpretation.

Consider souls. The Bible has many references to the soul. There have been philosophical and scientific challenges to belief in souls as spiritual substances that can live both inside of and outside of bodies. If we still take the biblical references to indicate belief in not only the reality but also the immortality of souls, how does belief in the immortality of the soul fit with strong biblical references to the resurrection of the body? I do not find a clear biblical teaching on this question. Perhaps Paul tried to express a relationship with his mention of a "spiritual body" in 1 Corinthians 15.44, but that term is itself puzzling. What kind of body would that be? Is it something that is simply beyond our comprehension? How exactly, if we may ask, are immortality of the soul and resurrection of the body supposed to be related in God's plans? I have not found a satisfactory answer to this question.

Consider also the conditions mentioned or at least suggested in the Bible for the next life. The Bible has references to heaven and also to Sheol, Hades, Gehenna, and Tarturus. There can be little objection to the concept of heaven as a wonderful provision made by God, even if some details are not entirely clear. But there is no explanation attempted in the Bible of the relationship between the various other terms. Especially puzzling is the relationship between Sheol and Gehenna, which appear to designate different conditions. Sheol (similar to Hades) suggests a place or condition for the spirits of all of the dead. Gehenna, at least with a literal and eternal interpretation, indicates a place or condition of punishment for some but not others. How do Sheol and Gehenna both fit into God's overall plan? There may be a good answer, but the Bible does not specifically tell us.

If we concentrate on biblical references to Gehenna, there have been various interpretations, including historical, eternal, literal, and symbolic. The traditional interpretation of Gehenna or hell as eternal agony planned by God is especially difficult. In my view, this interpretation appears to conflict with biblical teachings about God's love and God's justice.

I do not understand how belief in eternal agony can be compatible with belief in God's infinite love. The traditional justification for eternal agony is unforgiven offense against God's infinite holiness. I have a hard time believing that God's holiness or anything else exceeds God's love in requiring eternal agony. The standard view is that God shows love by providing a way for sinners to be forgiven, but what happens to those who are not redeemed? Eternal agony for some makes God look hateful and harsh, even sadistic, certainly not loving. Eternal agony would be greater cruelty than the most barbaric human could impose. Punishment of some kind may be compatible with God's love. Eternal agony does not appear to be.

Do we not have to consider God's justice along with God's love? God's justice may indeed require punishment for sin. We are all imperfect and fall short of the glory of God. We all do things we should not do and do not do things we should do. Perhaps we all deserve some kind of punishment. Yet the belief that God favors eternal agony for anyone makes the proposed punishment much worse than the crime, even for the worst people. Justice as fairness requires the appropriate amount of punishment, not too little and not too much. Eternal agony appears to be unjust in the sense of being greatly excessive in relation to the offense, even if we acknowledge that sin is against God as holy and infinite.

I realize that something such as eternal agony may be true even if we do not understand it or feel comfortable with it. I would still feel better about God's nature if unredeemed sinners who are denied heaven faced annihilation rather than eternal agony. Compared to heaven, annihilation sounds terrible. Compared to eternal agony, annihilation sounds merciful.

Is annihilation of the wicked an acceptable doctrine, perhaps with some punishment before extinction? The doctrine is not acceptable for those who emphasize the immortality of all souls. And annihilation does not seem to have much biblical support. But we should be careful. Even the beloved verse of John 3.16 does not say that those who believe in God's Son will not go to hell but that they will not perish. They will have everlasting life. We could get the idea that those who do not believe in God's Son will not have everlasting life, that is, will perish. Does that not sound like annihilation?

But what about the belief in the immortality of all souls? Can souls be annihilated? Even those who believe strongly in the reality of souls should remember a warning of Jesus. He said in Matthew 10.28 to "fear him who can destroy both soul and body in hell." I think the reference is to God according to the belief that nothing is impossible with God. If the soul represents life given by God, then God can end that life.

Speaking of conditional immortality may sound better than speaking of annihilation. Those who believe in God's Son will have everlasting life. This view suggests that eternal life is an added blessing. Those who do not believe in God's Son will not have everlasting life, unless God makes certain exceptions. It would be a shame not to have everlasting life. Yet missing eternal life sounds far better than the horrifying prospect of eternal agony.

I am not saying I know all God will do, because I do not. I believe God is ready to welcome everyone to heaven but allows the choice of declining his invitation. If it is possible for those not entering heaven to cease to exist, I believe that outcome would be much more in keeping with both God's love and justice than eternal agony would be.

While the Bible's teachings about God are important, even if not always completely clear, there is also the question of the reliability of the Bible. How much can we trust biblical teachings about God or anything? For those who think the Bible is God's inerrant Word, the Bible is entirely reliable, including whatever is said about God or anything else. I am among those who have concerns about the complete reliability of the Bible.

One concern about the Bible's reliability is the text. The original manuscripts of the books of the Bible no longer exist. There are copies of the original writings, but the copies do not always agree. Also, there may have been various errors by copyists. In spite of these difficulties, scholars believe they have come very close to establishing the original version of books of the Bible. But coming very close does not mean the combined text of copies of the Bible is exactly the same as the original manuscripts.

Another concern about the Bible's reliability is possible influence from other cultures on the writers of books of the Bible. For example, in 1872 George Smith reported to the Society of Biblical Archaeology in London that he had found an account of the Flood (the *Gilgamesh* epic) in ancient Assyrian tablets in the British Museum. Smith added in 1875 that he had also discovered a Babylonian account of creation (*enuma elish*).[1] Has some of the Bible, especially early sections of Genesis, been influenced by similar but older accounts? The implication is that the Bible may be due in part to views of other humans and may not have the special divine authority that many people attribute to it.[2]

But possible acquaintance with the views of others does not necessarily diminish the Bible's value. If there was indeed a massive flood in ancient times, would not people in various cultures have heard about it? Would not writers in various societies wish to give their versions of a great flood or the origin of the world? Whatever may be said about similar stories of different cultures, Christians may wish to believe the Bible is still the best written expression we have of beliefs about God.

Biblical reports of miracles can be a concern, at least with regard to the scientific reliability of the Bible. Scientists usually do not accept miracles. Some other people agree. Thus, accounts of miracles in the Bible can lead to rejection of the Bible either partly or entirely.

Those who believe in God as creator, as I do, generally have no difficulty with the possibility of miracles. The one who made heaven and earth can perform many wondrous signs. But some reports of miracles do seem strange. Miracles of healing may not seem especially strange because most of us probably know or have heard of very sick people who have had amazing recoveries. But how often have we talked to anyone who has spent three days inside of a big fish? How many times has there been a report that the sun "stood still"? While believing God can indeed perform miracles, we may wonder if all the biblical accounts of miracles should be taken literally. Allowing some symbolic interpretation of miracles would help views of the Bible's trustworthiness.

A further concern about the reliability of the Bible is the presence of inconsistencies. Clayton Sullivan has given many examples in *Toward a Mature Faith: Does Biblical Inerrancy Make Sense?* Sullivan gives 200 questions that challenge biblical inerrancy. Some of the questions have to do with what he views as questionable miracles or morally objectionable practices. But he does point out a large number of internally inconsistent statements in the Bible.

Consider some simple examples mentioned by Sullivan. Genesis 12.5-7 locates the oak of Moreh at Shechem. But Deuteronomy 11.29-31 locates the oak of Moreh at Gilgal.[3] According to 2 Samuel 24.9, David's census showed 800,000 warriors in Israel and 500,000 warriors in Judah. But according to 1 Chronicles 21.5, David's census showed 1,100,000 warriors in Israel and 470,000 warriors in Judah.[4] There are many other examples. If the Bible is inerrant, should there not be complete agreement?

While inconsistencies in the Bible are especially troublesome for those who wish to maintain belief in complete biblical inerrancy, they are not a huge problem for everyone. Most of the internal ones mentioned by Sullivan deal with comparatively minor details. The inconsistencies do not cover major teachings about God.

In spite of concerns about its complete reliability, the Bible remains a powerful source for thinking about God. Doubts about complete biblical reliability will probably continue, but so will faith in the major teachings of the Bible. I think the Bible should be taken very seriously but realize there can be many problems with interpreting the Bible. We should all learn, for example, from those people who mistakenly use the Bible to predict a date for the end of the world. We should be most careful with all interpretations of the Bible, including those concerning God.

When we turned to thinking about God and philosophy, we saw that philosophers usually attempt to think about God by reason apart from the Bible. Among many topics, philosophers have been concerned with trying to prove the existence of God and also with the nature of God.

Many philosophers have believed in God but have disagreed over whether the existence of God can be proved by human reason. Anselm, Thomas Aquinas, and William Paley all thought they had good arguments for God's existence. Pascal, Kierkegaard, and Tolstoy all believed in God but did not think God's existence could be proved by reason. Some philosophers, such as Camus, have not thought that God's existence could be proved because of their view that God is not real.

There are great difficulties with the various arguments for God's existence. I agree most with those philosophers who do not believe God's existence can be completely proved by reason. Consider the difficulties with various arguments.

Anselm's argument, for example, may look brilliant to some but has a quality of the absurd. How can a definition or a concept in the mind, however great, prove the existence of something outside of the mind? Existence does not occur simply from the thought that something has to exist. Also, Anselm did not have any content for his definition of God. There were no qualities or characteristics of that being than which nothing greater can be conceived. Exactly what was Anselm conceiving? We can agree that nothing can be conceived greater than God, but that thought by itself does not prove God's existence.

The arguments of Aquinas, in my opinion, are impressive but allow for a possible interpretation and not a necessary one concerning God's existence. Even if we go along with Aquinas and reject an infinite regress, we are not compelled to think there must exist a being who is the First Cause or the First Mover. An impersonal power as First Cause or First Mover is a reasonable possibility. As to necessary being, it would be the maximum of being, but the maximum of being would not have to be necessary being. If there is necessary being, how do we know apart from faith that it is God? Aquinas gets closer to what traditional theists consider to be God with the reference to the maximum of goodness. Believers identify God with the maximum of goodness, but reason does not require belief that goodness has to be divine. And there is no rational requirement that the maximum of being has to be the cause of all other degrees of being. The argument of Aquinas that implies an intelligent designer is close to the biblical concept of God as creator. But there are alternative explanations for what many interpret as designed. Thus I consider the arguments of Aquinas as establishing the possibility rather than the necessity of God's existence.

William Paley's argument for God's existence involved the world's having a purpose just as a watch has a purpose. Paley thought a watch's purpose of telling the time indicated a watchmaker who provided for that purpose. Does not the world show purpose and point to a maker of the world? But Paley did not establish a purpose for the world. Even thinking that some things in the world serve a purpose does not establish a purpose for the world itself. Paley's argument is clever but not convincing.

Might Paley's argument be strengthened by the claim that the world does have a purpose, which is to provide for and sustain life? There are many conditions that are necessary for life. The absence of one or more of these conditions or even slight modifications might make life impossible. We can believe that God is responsible for these conditions and for life itself. As a watch's purpose to tell the time indicates a watchmaker, does not the world's purpose to provide for life indicate a worldmaker (God)?

Although I believe this revised argument provides a strong indication for the possibility of God, I can see how the argument is not completely convincing for everyone. Might life have arisen apart from divine purpose and action? It seems impossible to believe that a watch could exist without a watchmaker, but many people believe the world and life could have come into existence without God.

Also, Anselm, Thomas Aquinas, and Paley concentrated on intellectual considerations that usually were impersonal. Anselm thought of God as that being than which nothing greater can be conceived, a correct idea for religious believers but an abstract concept when considered by itself. Aquinas, following Aristotle, thought of God in such terms as First Mover and First Cause and necessary being, not bad ideas but very intellectual rather than personal. Paley's argument for God's existence involved a watch, which was an intellectually respectable but mechanical rather than personal object.

In contrast to these views, the God of the Bible acts in history and in the individual lives of people. Perhaps Christian philosophers should do more than they have done in trying to prove the existence not of an impersonal power or principle but of the God of the Bible.

As I have indicated, I agree most with those philosophers who have believed in God even though they thought reason could not prove God's existence. I agree with some more than with others.

In spite of his great appeal for many, I am not convinced by the views of Kierkegaard concerning belief in God. He advocated giving up arguments for God's existence and making instead a leap of faith. He probably was reacting against what he considered to be impersonal intellectual concerns. His recommendation of a leap involved the thrill of a dedicated commitment. But he was not clear, and perhaps did not intend to be, on why a leap should be made. Even if a leap is to be made, why might it not be made toward doubt or even disbelief? I do not regard Kierkegaard's views on belief in God as persuasive.

I like the views of those thinkers who did not believe God's existence could be proved by reason but saw practical value in believing. Blaise Pascal claimed there was an even chance rationally of being right and that there would be great good from God if the person was correct. Leo Tolstoy thought that faith in God was against reason but that faith made it possible to live rather than want to die. Tolstoy thought faith in God gave life meaning. Although Pascal and Tolstoy did not think reason could prove God's existence, they still had reasons to believe. I like the thinking of Pascal and Tolstoy.

If we consider philosophers today, reliable numbers are not available concerning how many believe in God. There does seem to have been some change historically. If we look at the history of Western philosophy, most of the philosophers of the Middle Ages and even later were Christians. They believed strongly in God and often thought philosophic considerations could be used to support their belief. In modern times many philosophers have a secular viewpoint. There appear to be significant numbers of both theists and atheists among modern philosophers. Christian philosophers are most likely to be found at universities with a religious affiliation or heritage. Secular philosophers are most likely to be found at public universities and some private universities.

On the subject of the nature of God, we have considered various views among theistic philosophers. Traditional theism includes the idea of God as the all-powerful, all-knowing, and all-present creator who is good and loving. Traditional theists have the understanding that God is both above the world and within the world. Pantheism is the view that God is all or that everything is divine, that God and the world are the same. The idea of God as limited is the belief that God is good but not all-powerful and cannot completely eliminate evil. Some philosophers have claimed that God is not transcendent (above the world) but is completely immanent (within the world) in some way.

On the basic nature of God, I consider myself to be a traditional theist. I can understand the discouragement with evil that leads some to think God may be limited, but I disagree with that conclusion. I agree to some extent with those who think of God as within the world, but I believe God is also above the world. I believe but do not understand how God is everywhere.

The conviction in traditional theism of God's goodness has been both challenged and defended by philosophers. David Hume has questioned God's goodness because of conditions that allow for a great amount of suffering in the world. Hume thought God might be good but that the conditions would not make us think of a good God. John Hick has defended both God's goodness and sovereignty. Hick claimed a partial answer for evil by saying God's good goal of spiritual growth and development for people requires some suffering. Hick did not believe he had the total answer.

On the goodness of God I agree more with Hick than with Hume. I do agree with Hume that we have the ability to feel pain as well as pleasure, that situations in life often make us feel pain, and that we have limited ability to avoid painful situations. I do not agree with Hume that extremes of wind, temperature, and rain indicate inaccurate workmanship in creation. I agree with Hick that God can use at least some suffering for the good purpose of helping us to grow spiritually. I do wonder about what the explanation might be for other suffering, especially when suffering is intense, prolonged, or widespread, or perhaps all of these. Dostoevski's story of the little boy killed by the dogs illustrates the difficulty in finding spiritual growth in every situation of suffering. I also agree with Hick that hope for a better life in heaven strengthens belief in God's goodness.

Thinking about God and science was our third way for thinking about God. This kind of thinking is especially difficult because of widely different approaches. Many religious interpreters believe God has revealed truth, including scientific truth, in the Bible. Scientists generally do not think of the Bible as a good foundation for scientific knowledge. They emphasize gaining knowledge through reasoning about observation of the world, including observation enhanced by various kinds of instruments. These different approaches have led religious interpreters and scientific interpreters to some widely different conclusions about topics such as time, life, and space and about God.

As we think of time, a major question is how much time has passed since the beginning of the universe. According to Bishop Ussher's famous interpretation, God created the world in 4004 B.C. This claim was based on a literal interpretation of the Bible, including the view that there were six 24-hour days of creation. In disagreement, the prevailing scientific view today, based on various astronomical observations, is that the universe began 13.7 billion years ago with a Big Bang. Scientific accounts of the beginning of the universe usually do not mention God.

I do not believe the account of creation in Genesis 1 has to be interpreted literally with regard to time. There is, among other questions, the meaning of "day" in the story before the creation of the sun on the fourth day. How could there be a "day" of 24 hours (or perhaps any definite length) without the sun? I do accept the biblical view of God as creator but believe scientists may be correct in thinking that the universe began 13.7 billion years ago. Even though scientific calculations may be adjusted, it is likely the universe is extremely old.

When we consider God and life, a strong religious interpretation based on Genesis 1 and 2 is that God created various life forms or species, including humans, much as they are today. Many people are convinced there has not been significant change in species since creation.

Charles Darwin's studies led him to reject the idea of the immutability or changelessness of species. Darwin thought life began long ago with tiny organisms in the sea and then developed into various forms, including mammals, Old World monkeys, and eventually humans. Darwin's writings include a statement that is favorable toward a creator, but Darwin did not believe in the direct creation of humans by God.

I do not have the scientific knowledge that would allow an adequate evaluation of Darwin's views. However, I have no difficulty with thinking that God may have developed life, including human life, over a long period of time. I do believe that scientists should improve their explanations of evidence for their views on when, where, how, and why life began, animal life and plant life became separate, gender appeared, and human life came into existence.

I think God is a better explanation for the origin of life than believing life "just happened" or conditions were fortunately favorable for life to begin. I continue to believe God is responsible for the beginning of life, including human life, whenever and however God may have done it. I believe the creation stories in Genesis may be interpreted as poetic narratives that mainly express belief that God is responsible for the origin of life.

Thinking about God and space may involve the biggest scientific challenge for religious interpreters, greater than how long ago the universe began or when and how life began. There is the situation of the immense size of the universe, of the extent of occupied space.

Scientists have indisputably established that the universe is enormous. We generally have accepted that there are many millions of miles between our planet and the sun. Even more impressive, there are reports not of miles but of billions of light-years between some objects in space. Remember that light travels at approximately 186,000 miles a second, so the distance light travels in a year is gigantic. With the thought that this already huge universe is expanding and even expanding at a rapid rate, we might wonder how big the universe can get. And what is beyond space, beyond the occupied space of the known universe? The possibilities are staggering.

How should we understand God in relation to the size of the universe? How should we understand the significance of humans in relation to the size of the universe? Religious interpreters have given comparatively little attention to these questions. Modern scientific interpreters of space have expressed little interest in God. Their interest is mostly in whether or not there is intelligent life beyond our planet.

When we think about God and space, there is almost no debate about what scientists have discovered concerning the vast size of the universe. Simply looking at the sky, especially the night sky with moon and stars, gives us some sense of the immensity of space. But what are the implications?

I believe the greatness of the universe is an indication of the greatness of God as creator, far more than what many of us may have thought. We may have difficulty in expressing the "place" of God regarding space but still believe in God's grandeur.

How should we understand the size of the universe in relation to our significance and God's interest in us? With the universe as large as it is now and as large as it may become, why should we think that God cares about the human inhabitants of a tiny globe in the vastness of space? In the days when it was thought that the sun revolved around our planet, people could claim a special significance. Now we understand our whole planet to be barely a speck in our galaxy, much less in the whole universe.

But there are assurances that humans still are important. As we have considered, we have assurances from the Bible and from the claims of people who believe God is active in their lives. We know that some comparatively small things have greater significance than their size would suggest. And there is the apparently unique ability of humans to think.

Although we do have various assurances of our significance in the universe and in God's sight, we should be careful. Do we ask, even demand, special attention for ourselves without considering our smallness in the huge expanse of the universe? Have we overestimated our importance for God? We should probably express our faith about God's concern for us with hopeful humility rather than with complacency, dogmatism, or arrogance.

If we look at the overall relationship between religion and science, there are some challenges. Consider Andrew White's *A History of the Warfare of Science with Theology in Christendom*. Although published at the end of the nineteenth century, the two volumes show the advance of science over many religious views that now seem foolish (for example, comets as divine signs) or even cruel (such as witch trials). Are there additional religious views that should be adjusted or abandoned?

The biggest scientific challenge for religion today concerns God. Many scientists are not favorable toward various religious interpretations and also do not believe in God. An example is Richard Dawkins, who strongly disagrees with religious apologists who try to find theists among admired scientists. Hawkins not only thinks the majority of British and American scientists are atheists, but he also believes that such exceptions to atheism as Francis Collins provide "amused bafflement" to their scientific colleagues. Dawkins also suggests that atheism tends to increase in direct proportion to the status of scientists.[5]

Some support for the views of Dawkins can be found in a special report published in 2009 by the Pew Research Center for the People and the

Press along with the American Academy for the Advancement of Science. The random sampling includes the 2,533 members of the scientific association and 2,001 U.S. adults from the general population. Of the scientists surveyed, only a third said they believed in God.[6]

I think the scientists are probably correct in many of their ideas about time, life, and space. Whatever the exact number of years, the universe probably began much earlier than 10,000 years ago. Life also likely began a long time ago and has gone through many stages of development. The vastness of the universe, of occupied space, is staggering.

I do not think these various scientific views have to lead to disbelief in God. Biblical writers, many philosophers, and even some scientists have believed in God. Also, I think God is the explanation for the universe. As a further thought, I am impressed by Pascal's explanation of how it is better to believe in God than not to believe.

I do think scientific claims should lead us to some adjustments in our views of God. We should not interpret everything in the Bible about God literally. And we should allow for the possibility that God is far greater than we have sometimes imagined.

I continue to believe not only that God is real but also that God cares for us. I believe there is no one better to follow concerning God than Jesus, who was the best expression we have had of love for God and love by God. Those of us who believe in God have many challenges but also have a magnificent conviction and a glorious hope.

What do you think about God? What are your beliefs?

Notes

[1] See Richard Briggs' article on "The Hermeneutics of Reading Genesis after Darwin" in *Reading Genesis after Darwin*, ed. Stephen C. Barton and David Wilkinson (New York: Oxford University Press, 2009), 59.

[2] This view was expressed, for example, by Andrew White at the end of the nineteenth century. See chapter 20 on "From the Divine Oracles to the Higher Criticism" in Andrew White, *A History of the Warfare of Science with Theology in Christendom* (Amherst, NY: Prometheus Books, 1993).

[3] Clayton Sullivan, *Toward a Mature Faith: Does Biblical Inerrancy Make Sense?* (Decatur, GA: SBC Today, 1990), 70.

[4] Ibid., 81.

[5] See Richard Dawkins, *The God Delusion* (Boston: Houghton Mifflin Co./Mariner Books, 2008), 120-130.

[6] Religion News Service, "Report finds one-third of scientists believe in God," *Baptists Today*, September 2009, 22.

BIBLIOGRAPHY

Attridge, Harold W., ed. *The Religion and Science Debate: Why Does It Continue?* New Haven: Yale University Press, 2009.

Barton, Stephen C. and David Wilkinson, eds. *Reading Genesis after Darwin.* New York: Oxford University Press, 2009.

Bell, Rob. *Love Wins.* New York: HarperOne, 2010.

Calvin, John. *Institutes of the Christian Religion.* Translated by Henry Beveridge. 2 vols. London: James Clarke & Co., 1953.

Catechism of the Catholic Church. English translation. Liguori, MO: Liguori Publications, 1994.

Collins, Francis S. *The Language of God.* New York: Free Press, 2006.

Crockett, William, ed. *Four Views on Hell.* Grand Rapids, Zondervan, 1996.

Darwin, Charles. *The Descent of Man.* 2nd ed., John Murray, 1879. London: Penguin Books, 2004.

Darwin, Charles. *The Origin of Species.* Originally published in 1859. Edison, NJ: Castle Books, 2004.

Dawkins, Richard. *The God Delusion.* Boston: Houghton Mifflin Company, Mariner Books, 2008.

Dawkins, Richard. *The Greatest Show on Earth: The Evidence for Evolution.* New York: Free Press, 2009.

Gould, James A., ed. *Classic Philosophical Questions.* 8th ed.. Englewood Cliffs, NJ: Prentice-Hall, Inc., 1995.

Hawking, Stephen. *A Brief History of Time.* 10th anniv. ed. New York: Bantam Books, 1998.

Hawking, Stephen. *Stephen Hawking's A Brief History of Time: A Reader's Companion.* New York: Bantam Books, 1992.

Hawking, Stephen. *The Universe in a Nutshell.* New York: Bantam Books, 2001.

Hick, John. *Philosophy of Religion.* 2nd ed.. Englewood Cliffs, NJ: Prentice-Hall, Ind., 1973.

Honer, Stanley M., Thomas C. Hunt, and Dennis L. Okholm. *Invitation to Philosophy.* 7th ed.. Belmont, CA: Wadsworth Publishing Company, 1996.

L'OSSERVATORE ROMANO. English ed. November 4, 1992.

Ludwig, Theodore M. *The Sacred Paths.* 2nd ed. Upper Saddle River, NJ: Prentice Hall, 1996.

Plato. *Great Dialogues of Plato.* Translated by W. H. D. Rouse. Edited by Eric H. Warmington and Philip G. Rouse. New York: The New American Library, 1956.

Peters, Ted and Martinez Hewlett. *Can You Believe in God and Evolution?* Nashville: Abingdon Press, 2006.

Sagan, Carl. *Cosmos.* New York: Ballantine Books, 1985.

Sullivan, Clayton. *Toward a Mature Faith: Does Biblical Inerrancy Make Sense?* Decatur, GA: SBC Today, 1990.

Tyson, Neil deGrasse and Donald Goldsmith. *Origins.* New York: W. W. Norton & Company, 2004.

White, Andrew D. *A History of the Warfare of Science with Theology in Christendom.* 2 vols. 1st ed., New York: Appleton, 1896. Buffalo, NY: Prometheus Books, 1993.

Wright, G. Ernest and Reginald H. Fuller. *The Book of the Acts of God.* Garden City, NY: Doubleday & Company, Inc., 1957.

CPSIA information can be obtained at www.ICGtesting.com
Printed in the USA
LVOW13s1800120913

352156LV00003B/8/P